CHINA IN THE 21ST CENTURY

GOING GREEN IN CHINA

POLICY AND POTENTIAL

CHINA IN THE 21ST CENTURY

Additional books in this series can be found on Nova's website under the Series tab.

Additional E-books in this series can be found on Nova's website under the E-books tab.

ENVIRONMENTAL SCIENCE, ENGINEERING AND TECHNOLOGY

Additional books in this series can be found on Nova's website under the Series tab.

Additional E-books in this series can be found on Nova's website under the E-books tab.

CHINA IN THE 21ST CENTURY

GOING GREEN IN CHINA

POLICY AND POTENTIAL

SPENCER B. TALSCOTT
EDITOR

Nova Science Publishers, Inc.
New York

For permission to use material from this book please contact us:
Telephone 631-231-7269; Fax 631-231-8175
Web Site: http://www.novapublishers.com

Library of Congress Cataloging-in-Publication Data

Going green in China : policy and potential / Spencer B. Talscott, editor.
p. cm.
Includes index.
ISBN 978-1-62100-691-6 (hardcover)
1. Environmental policy--China. 2. Environmental protection--China. 3. China--Environmental conditions. I. Talscott, Spencer B.
GE190.C6G65 2011
333.720951--dc23

2011037417

Published by Nova Science Publishers, Inc. † New York

CONTENTS

PREFACE

This book examines China's greenhouse gas emissions and mitigation policies. Debate on potential climate change legislation in the United States has been influenced by China's surging GHG emissions, and uncertainty over whether, how, and when China might alter that trend. There is concern that strong U.S. domestic action taken without Chinese reciprocity would unfairly advantage China in global trade, and fail to slow significantly the growth of atmospheric concentrations of GHGs. The governments of both China and the United States have indicated some closure of their gap on future actions to address climate change by agreeing on national pledges to GHG targets and mitigation actions rather than binding international obligations. China is also engaged with many other countries in bilateral programs to build its governance and technological capacities to abate its GHG emissions.

Chapter 1- The 112th Congress continues to debate whether and how the United States should address climate change. Most often, this debate includes concerns about the effects of U.S. greenhouse gas (GHG) emissions controls if China and other major countries were not to take comparable actions. China recently surpassed the United States to become the largest emitter of human-related GHG globally, and together, the two nations emit about 40% of the global total (with shares of 21% and 19% respectively).

China's GHG emissions are growing rapidly and, even with policies adopted by China, are expected to rise until at least 2030. The emissions growth is driven by China's rapid economic and industrial growth and its reliance on fossil fuels despite measures to raise the shares of non-fossil energy sources. China requires 50% more energy to produce one billion dollars of GDP (its "energy intensity") compared with the United States. Over the past two decades, strong government directives and investments have dramatically reduced the energy and GHG intensities of China's economy, though the rates of improvement leveled off in the 2000s, and even reversed in subsequent years. A renewed emphasis on improving energy and GHG intensity emerged in the 11th 5-Year Plan, from 2006-2010, and the government says the nation nearly achieved its aggressive goal to reduce by 20% the energy required to produce GDP. In the context of China's 12th 5-Year Plan, from 2011-2015, leaders have set targets to further reduce energy intensity by 16% by 2015. Along with measures to reduce pollution and increase the shares of non-fossil fuels in the energy sector, China has set goals to improve its CO_2 intensity by 40-45% by 2020, with an interim target in the 12th 5-Year Plan of 17% by 2015. Even if these targets are achieved, China's GHG emissions are expected to rise in absolute terms. In addition, the frequency, transparency, and data quality of China's reporting of its GHG emissions and mitigation actions (including underlying energy and other data) have been a challenging diplomatic issue between the United States and China and in the climate change negotiations. China has resisted reporting and reviews comparable to what other industrialized nations or what many developing countries accept. While technical bilateral cooperation on data has been productive and China has moved politically toward better information sharing, the continuing lack of transparency is apparent in uncertain emissions estimates and projections.

Chapter 2- Local level actions are crucial for achieving energy-saving and greenhouse gas emission reduction targets. Yet it is challenging to implement new policies and actions due to a lack of information, funding, and capacity. This is particularly the case in developing countries such as China. Even though national energy intensity and carbon intensity targets have been set, most local governments do not have the knowledge regarding actions to achieve the targets, the cost-effectiveness of policies, the possible impact of policies, or how to design and implement a climate action plan.

This paper describes a guidebook that was developed to motivate and provide local governments in China with information to create an action plan to tackle climate change and increase energy efficiency. It provides a simple step-by-step description of how action plans can be established and essential elements to be included—from preparing a GHG emission inventory to implementation of the plan. The guidebook also provides a comprehensive list of successful policies and best practices found internationally and in China to encourage low carbon development in industry, buildings, transportation, electric power generation, agriculture and forestry.

This paper also presents indicators that can be used to define low-carbon development, as well as to evaluate the effectiveness of actions taken at an aggregated (city) level, and at a sectoral or end use level. The guidebook can also be used for low carbon development by local governments in other developing countries.

Chapter 3- The dominant image of China's energy system is of billowing smokestacks from the combustion of coal. More heavily dependent on coal than any other major country, China uses it for about 70 percent of its energy (NBS, 2008). Furthermore, until recently, China had very few environmental controls on emissions from coal combustion; recent efforts to control sulfur dioxide (SO_2) emissions appear to be meeting with some success (Economy, 2007, 2009).

Chapter 4- China has been, is, and will continue to be a coal-powered economy. In 2007 Chinese coal production contained more energy than total Middle Eastern oil production. The rapid growth of coal demand after 2001 created supply strains and bottlenecks that raise questions about sustainability.

Urbanization, heavy industrial growth, and increasing per-capita income are the primary interrelated drivers of rising coal usage.

In 2007, the power sector, iron and steel, and cement production accounted for 66% of coal consumption.

Power generation is becoming more efficient, but even extensive roll-out of the highest efficiency units would save only 14% of projected 2025 coal demand for the power sector.

A new wedge of future coal consumption is likely to come from the burgeoning coal-liquefaction and chemicals industries. If coal to chemicals capacity reaches 70 million tonnes and coal-to-liquids capacity reaches 60 million tonnes, coal feedstock requirements would add an additional 450 million tonnes by 2025.

Even with more efficient growth among these drivers, China's annual coal demand is expected to reach 3.9 to 4.3 billion tonnes by 2025.

In: Going Green in China: Policy and Potential ISBN: 978-1- 62100-691-6
Editors: Spencer B. Talscott

Chapter 1

CHINA'S GREENHOUSE GAS EMISSIONS AND MITIGATION POLICIES*

Jane A. Leggett

SUMMARY

The 112th Congress continues to debate whether and how the United States should address climate change. Most often, this debate includes concerns about the effects of U.S. greenhouse gas (GHG) emissions controls if China and other major countries were not to take comparable actions. China recently surpassed the United States to become the largest emitter of human-related GHG globally, and together, the two nations emit about 40% of the global total (with shares of 21% and 19% respectively).

China's GHG emissions are growing rapidly and, even with policies adopted by China, are expected to rise until at least 2030. The emissions growth is driven by China's rapid economic and industrial growth and its reliance on fossil fuels despite measures to raise the shares of non-fossil energy sources. China requires 50% more energy to produce one billion dollars of GDP (its "energy intensity") compared with the United States. Over the past two decades, strong government directives and investments

---

* This is an edited, reformatted and augmented version of a Congressional Research Service publication, CRS Report for Congress R41919, from www.crs.gov, dated July 18, 2011.

have dramatically reduced the energy and GHG intensities of China's economy, though the rates of improvement leveled off in the 2000s, and even reversed in subsequent years. A renewed emphasis on improving energy and GHG intensity emerged in the 11th 5-Year Plan, from 2006-2010, and the government says the nation nearly achieved its aggressive goal to reduce by 20% the energy required to produce GDP. In the context of China's 12th 5-Year Plan, from 2011-2015, leaders have set targets to further reduce energy intensity by 16% by 2015. Along with measures to reduce pollution and increase the shares of non-fossil fuels in the energy sector, China has set goals to improve its CO_2 intensity by 40-45% by 2020, with an interim target in the 12th 5-Year Plan of 17% by 2015. Even if these targets are achieved, China's GHG emissions are expected to rise in absolute terms. In addition, the frequency, transparency, and data quality of China's reporting of its GHG emissions and mitigation actions (including underlying energy and other data) have been a challenging diplomatic issue between the United States and China and in the climate change negotiations. China has resisted reporting and reviews comparable to what other industrialized nations or what many developing countries accept. While technical bilateral cooperation on data has been productive and China has moved politically toward better information sharing, the continuing lack of transparency is apparent in uncertain emissions estimates and projections.

Chinese negotiators adhere to the principle of "common but differentiated" responsibilities, agreed in the United Nations Framework Convention on Climate Change (1992). They argue that emissions per person in China are low, that raising incomes must be their highest priority, and that industrialized countries bear primary responsibility for the historical buildup of GHGs in the atmosphere; therefore the industrialized countries should lead in mitigating emissions domestically. Industrialized countries also, they say, should assist developing countries with financial and technological support to mitigate emissions and adapt to coming change.

Debate on potential climate change legislation in the United States has been influenced by China's surging GHG emissions, and uncertainty over whether, how, and when China might alter that trend. There is concern that strong U.S. domestic action taken without Chinese reciprocity would unfairly advantage China in global trade, and fail to slow significantly the growth of atmospheric concentrations of GHGs. The governments of both China and the United States have indicated some closure of their gap on future actions to address climate change by agreeing on national pledges to GHG targets and mitigation actions rather than binding international obligations. China is also engaged with many other countries in bilateral programs to build its governance and technological capacities to abate its GHG emissions.

WHY CHINA'S GREENHOUSE GAS EMISSIONS MAY MATTER

China's greenhouse gas (GHG) emissions have drawn attention in the United States because of their environmental and economic implications. China's actions to address climate change also hold implications for broader economic and security concerns in the United States.

Scientific evidence that the Earth's climate is changing, and that human-related GHG emissions are a major driver of that change, has led to debate over whether and how to control GHG emissions. Once emitted, GHG persist in the atmosphere for years to centuries (and for some gases, millennia). They allow solar radiation to enter the Earth' system, but prevent much of the absorbed energy from escaping back out to space. Scientists agree that the Earth's atmosphere serves as a "blanket" that warms the Earth's surface and that a certain concentration of GHG is essential to maintain the planet at habitable temperatures. There is less agreement on how much warming would result from the higher atmospheric GHG concentrations expected if emissions from fossil fuel use, deforestation, and some agricultural and industrial processes continue. Scientific concerns in the 1980s led to initiation of inter-governmental discussions in 1989 to stabilize GHG concentrations and avoid potentially "dangerous" global temperature rise. These concerns led to negotiation of the United Nations Framework Convention on Climate Change (UNFCCC).

In the late 1980s, climate experts broadly understood that climate change driven by human-related GHG emissions was a global challenge: all major emitting countries would need to engage in slowing then reducing their emissions of GHG as well as increasing GHG removals by "sinks" (e.g., growing forests). When the UNFCCC was opened for signature in 1992,[1] the already industrialized countries[2] emitted almost 80% of the global carbon dioxide (CO_2) from energy and industry.[3,4] The CO2 emissions of the United States and the European Union were about 23% and 20%, respectively, of the global total. China's were about 11%. All the "developing" countries at the time contributed about one-third. Low income countries saw GHG-driven climate change as a problem made by the industrialized countries. Considering low income countries' challenged financial, technological, and governance capacities, they were not included in the UNFCCC's Annex I, which lists countries with quantitative GHG control targets for the 1990s. Nonetheless, the UNFCCC contained a principle of "common but differentiated responsibilities" among its Parties, with consensus that the already industrialized countries should lead in controlling their emissions and that all countries have obligations to address climate change. Annex I established a bifurcation between the Parties listed in Annex I and the Non-Annex I Parties. (Countries are frequently referred to as "developed" versus "developing," although the distinction is undefined and arguably a misleading simplification of the spectrum of differences among countries).

Scientific analyses have concluded that rising GHG concentrations in the atmosphere cannot be stabilized unless all major emitting countries abate their net emissions to near zero.[5] Despite efforts of many countries to reduce their GHG emissions, the continued and rapid growth of emissions from such large emitters as China and the United States has called into question the efficacy of the UNFCCC in meeting its objective of stabilizing concentrations of GHG in the atmosphere. As China's share of global GHG emissions has grown from about 11% in 1990 to about 21% today, and continues to grow, a broad set of observers have concluded that effectively slowing human-induced climate change depends on Chinese reductions of its emissions, as well as reductions from the United States and all other large emitters.

Issues for Congress

U.S. congressional debate on potential climate change policies in the United States frequently invokes China's (and other emerging economies') surging greenhouse gas (GHG) emissions as well as skepticism over whether, how, and when China might alter that trend. Some are concerned that U.S. investment in GHG controls without comparable Chinese actions would unfairly advantage China in global trade, and fail to slow human-induced climate change. In contrast, others point to China's announced aggressive actions to reduce GHG emissions by deploying efficient and non-fossil fuel technologies, and warn that U.S. businesses could fall technologically and competitively behind China in the energy sector and international trade. Moreover, some suggest that Chinese policies may help buffer their production from rising and volatile fossil fuel prices, while the United States' production costs could remain subject to fluctuating world energy prices.

Some experts point to the connections of China's GHG control policies to broader congressional concerns: China's role in world markets for energy, forest products, rare earth metals, and other strategic resources. Some analysts have noted that potential impacts of climate change on China's ecosystems, agriculture, water resources, and coastal zones[6] could affect that nation's economic and political prospects, with both economic and security risks for the United States.[7]

This report lays a groundwork for congressional consideration of China's GHG policies. It describes the underlying economic and environmental context for China's GHG emissions then provides the magnitude and uncertainties of available estimates. It discusses the factors driving China's emissions and summarizes some of the best described elements of China's strategy to mitigate its GHG emissions. A brief section identifies key points on China's status in international cooperation.

China's Economic and Environmental Context

Over the past decade, the Chinese people and its leadership have come to view the country's economic well-being and its environmental quality as inextricably correlated. Figures for Gross Domestic Product without accounting for losses to air and water quality, human health, etc. exaggerated actual improvements in living standards. This has resulted in a shift to "harmonious growth" that tries to balance GDP growth with improving living standards.

Between 1980 and 2009 the Chinese economy grew at an average annual rate of 10% (growing by factor of 16 over the period).[8] In the past decade, China's annual growth rates varied between 8% and 14%, over which time its economy has tripled in size. China is now the second largest economy in the world, behind the United States. The country held $3.2 trillion in foreign exchange reserves by the June 2011.[9,10] Hundreds of millions of Chinese have improved their standards of living. These facts lead some to contend that China is no longer a developing country.

Still, a recent World Bank report estimates that up to 200 million people (out of 1.3 billion) in China continue in poverty, living on less than $1.25 a day in 2005.[11] The Chinese government perceives that its legitimacy depends on continuing to improve the living standards of its population. Its economic policies are stated to be strongly aimed at eliminating poverty and raising average incomes. China's achievements so far are reflected in Gross Domestic Product (GDP) per capita in 2009 of $6,800 in international dollars (int$),[12] up 22% from 2007 and up 289% from 2000, according to the World Bank's

World Development Indicators. Still, per capita incomes in China were about one-eighth those of Americans levels (respectively, int$5,594 and int $46,628).[13]

Table 1 provides selected economic, energy, and greenhouse gas statistics for China and the United States. In 2009, China's population was more than four times that of the United States, while its economy, as measured using "purchasing power parities,"[14] was just over half as large. Even with relatively slow population growth (0.5% annually) and high projected economic growth rates of at least 7% annually, it likely would take many years for the living standards of average Chinese to reach those of average Americans.

China's national plans call for structural shifts in its economy. Much of China's recent growth occurred in industrial production, especially of goods for export. In the past few years, China's policies have attempted to shift from emphasizing heavy industry (such as steel) and production for export to stimulating domestic consumption and meeting domestic demands for improved quality of life. China has reduced its incentives for export-oriented production and increased incentives for investments in higher value-added goods and services. These policies, combined with increasing economic integration in Southeast Asia, could decrease the share of heavy manufacturing in China's economy, although preliminary data from early 2011 show that heavy manufacturing remains strong.

Historically, China's promotion of monetary economic growth has led to severe environmental degradation. Many Chinese policymakers appear now to realize, however, that they cannot sustain an "unbalanced" approach that emphasizes industrialization at any cost, especially not in the developed regions of the country. Environmental pollution has become so bad in places that social and political stability are at risk. Officially recognized "public order disturbances" grew from 58,000 in 2003 to 87,000 in 2005, many due to environmental pollution and land-takings stemming from government corruption.[15] Chinese officials have indicated that they seek a "harmonious society" that would entail slower GDP growth with less environmental degradation.

In 2007, the World Bank, working with the Chinese government, estimated that the cost of outdoor air and water pollution to China's economy totaled around US$100 billion annually, or 5.8% of China's GDP.[16] In other words, if non-monetized losses to China's resource assets (e.g., clean water, etc.) were netted from the current (financial) accounts, GDP would have been 5.8% lower. Related to such findings, the Chinese government raised the stated priority of environmental protection in its 11th Five Year Plan (2006-2010).

Table 1. Selected Statistics for China and the United States (data are for 2009 unless otherwise noted)

	China	United States
Population (millions)a	1,331	307
Population Growth (annual %)	0.5	0.9
Gross Domestic Product using Purchasing Power Parities[a] (GDPppp in billion international $)	9,091	14,119
GDP per capita (using international $)	6,828	45,989
GDP growth (%)	14.2	1.9
Primary energy consumption (million tons oil equivalent MTOE)	2,188	2,204
Energy Use per Capita (MTOE oil equivalent per capita)[c]	1.6	7.2
Energy Intensity (MTOE used per int$ millions of GDPppp)	235	156
Electricity Consumption per Capita (kWh per capita)	2,791	13,506
Carbon Dioxide Emissions (1,000 metric tons CO_2) in 2009	7,547	5,904
Greenhouse Gas Emissions (1,000 metric tons $_{CO2e}$)[b] in 2005	7,527	7,282
Greenhouse Gas Emissions per Capita (metric tons CO_2e per capita) in2005[d]	6	25
Greenhouse Gas Emissions Intensity in 2005 (tons CO_2e per int$1000 of Gross National Product, using PPPs)	1.4	0.6

Sources and Notes: Most economic data are available from the World Bank, *World Development Indicators*, for both China and the United States. BP provides energy data for 2009 for both countries. Non-CO_2 emissions data are available from the IEA most recently for 2005, and are unofficial for China. Therefore, indicators using six GHG are provided for the year 2005.

[a] World Bank Group, *World Development Indicators*, data extracted July 5, 2011.

[b] The term "CO2e," or "carbon dioxide equivalents," quantifies six greenhouse gases (CO2, CH4, N2O, SF6, HFC and PFC) according to a "Global Warming Potential" (GWP) index of their relative estimated effects on global warming over a 100-year period.

[c] British Petroleum (BP), Statistical Review of World Energy 2011 at http://www.bp.com/sectionbodycopy.do? categoryId=7500&contentId=7068481.

[d] International Energy Agency (IEA) greenhouse gas database (extracted January 8, 2008).

Chinese central government officials have over the past decade pursued a combination of measures to control air, water, and soil pollution, and state that they are striving to build a "recycling" industrialized economy[17] to ease environmental pressures. Their efforts have met with mixed success. Even when national officials genuinely want to encourage environmental improvement, local officials may seek to avoid enforcement of environmental regulations (and may not report accurate data) in order to maximize industrial growth and employment. As will be reflected later in this report, the degree to which national goals and measures will be achieved remains an important question.

Controlling local and regional pollutants like oxides of sulfur and nitrogen, particulates, and mercury has been difficult because of the difference in priorities of local and central government officials, as well as insufficient monitoring data and enforcement of national requirements. Controlling GHG emissions in China is even harder. For conventional pollutants (e.g., particulate matter or water contamination), both mitigation costs and impacts are local or regional; averaged nationally, polluting nations will largely pay the costs of that pollution either financially or in lower well-being. But with GHG emissions, mitigation costs may be local, while climate impacts are global. Without *shared* international action, this can lead to a "tragedy of the commons" phenomenon[18] where the shared resource is not adequately stewarded, or where some people take responsibility while others, who do not control their emissions, become "free riders" on the efforts of those who do.[19]

Over the past three years, China's leadership has demonstrated an increasing realization that it has ownership in the outcomes of a warming world. Increasingly, it recognizes that it too would bear potential climate change costs—of increasing storm intensity, rising sea levels, shifting water availability, challenged agricultural productivity, changing disease patterns, as well as other anticipated impacts.[20]

China's recently announced 12th 5-Year Plan, covering 2011-2015, says the nation should act to balance economic growth with environmental protection. How China reconciles domestic politics with international relations, and "fairness" arguments with pledges of actions to suppress its GHG emissions (to be discussed later) remains to be seen.

CHINA'S GREENHOUSE GAS EMISSIONS

Transparency Concerns

Data from China are equivocal. No one knows precisely the scale of China's GHG emissions or its removals of CO_2 from the atmosphere by vegetation. China has not reported its emissions estimates for any year later than 1994[21,22] although an unofficial estimate for 2004 was released (Figure 1).[23] (By comparison, the United States and most industrialized countries have been reporting officially and annually according to internationally agreed guidance since 1995.) The lack of China's reporting, transparency, and acceptance of international review of GHG emissions estimates (and underlying data) has been a major point of contention between China and the United States (and other countries) in the UNFCCC negotiations. While China may not have reliable information at this time, its insistence that reporting should be a point of "differentiation" among countries has not helped China convince others of its sincerity in undertaking domestic actions to slow its GHG emissions growth. China has alleged that international review of its emissions estimates could be "intrusive"; other countries, including the United States, already engage in such reviews, including in-country reviews and discussions with independent third-parties. Since the Copenhagen Accord in 2009, China has agreed to biannual reporting and "consultations," though the terms of those agreements are yet to be defined.

Available Estimates of China's GHG Emissions

The government is currently preparing its second ever GHG emissions inventory, expected to be released in 2012. Based on activity data of varying quality, estimated GHG emissions in China in 2005 were around 7-7.5 billion metric tons of CO_2-equivalent,[24] with CO_2 constituting 78-84% of the total. Methane (CH_4) emissions were around 11-13%, nitrous oxide (N_2O) about 1%, and the synthetic gases (sulfur hexafluoride—SF_6, perfluorocarbons—PFC, and hydrofluorocarbons—HFC) together less than 1%. (These are the six GHG covered by the Kyoto Protocol.)

The Chinese government estimates[25] that the country offset a portion of its GHG emissions with removals (sequestration) by forests: "from 1980 to 2005, a total of 3.0 billion tons of CO_2 were absorbed by afforestation, a total of 1.6 million tons of CO_2 were absorbed by forest management, and 0.430 million tons of CO_2 from deforestation were saved."[26] Removals of CO_2 from the atmosphere by land use, land use change and forestry are much more difficult to quantify in all countries than emissions by other human activities.

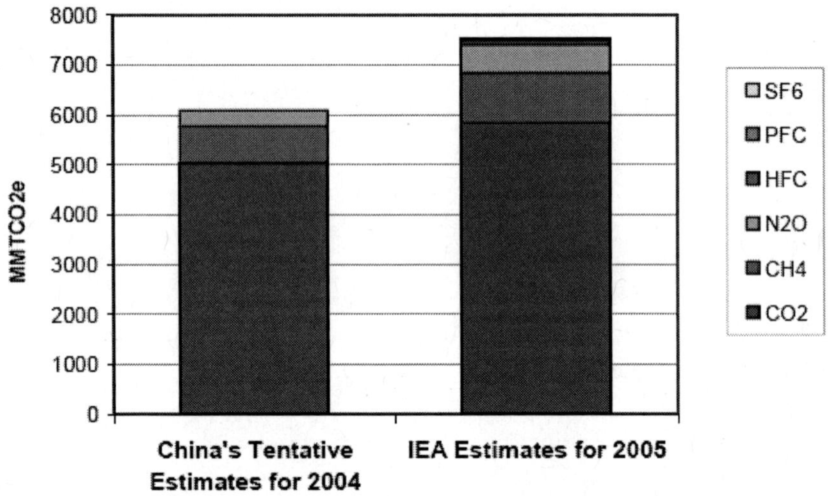

Source: CRS figure with estimates from Xinhua (June 4, 2008); IEA greenhouse gas database (extracted Jan. 8, 2008); World Resources Institute, Climate Analysis Indicators Tool (data extracted July 11, 2011); G.Q. Chen and Bo Zhang, "Greenhouse Gas Emissions in China 2007" *Energy Policy* Vol 38:10 pp. 6180-6193 (October 2010).

Notes: Few estimates of non-CO_2 GHG emissions in China are available, and those that exist use varying years of analysis, scopes, and methods. This figure shows two available estimates for different years, illustrating the similarities and uncertainties regarding China's GHG emissions. Even carbon dioxide emissions, which are straightforward to calculate from energy and cement data, are no better than the quality of the underlying energy data. Though the IEA estimates include emissions of PFC and SF6, the percentages are too small to appear in this figure.

Figure 1. Two Estimates of Recent Chinese GHG Emissions; (by gas, where data were available).

Figure 2 ranks the the IEA's estimates of the world's leading GHG emitters in 2005, the latest year for which non-CO_2 GHG data are available for many countries, including China. According to IEA's estimates for the six GHG, China and the United States were each responsible for about 17% of global GHG emissions in 2005. Using data for energy and cement CO_2 only,

China and the United States in 2009 emitted about 21% and 19%, respectively, of the global total.[27]

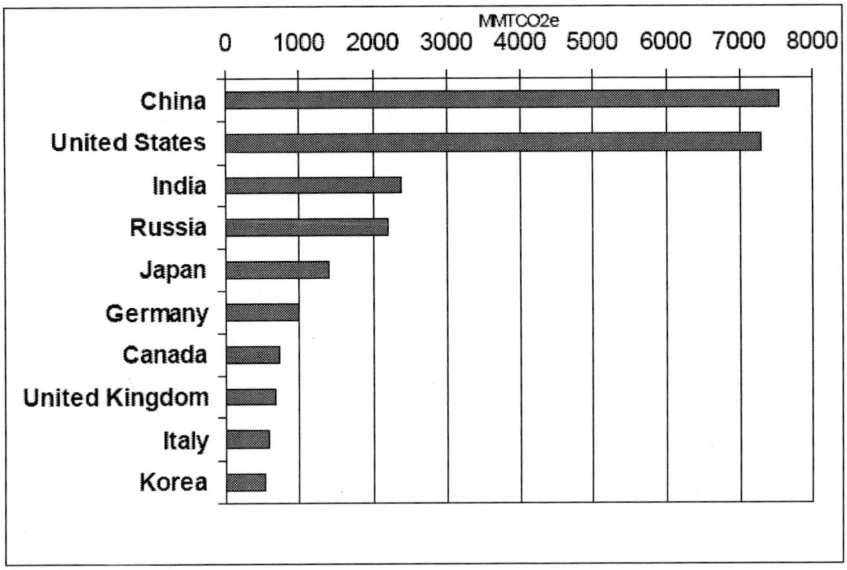

Source: CRS graphic from IEA estimates (extracted January 8, 2008).

Figure 2. Top GHG Emitters in 2005.

Recent Rates of Growth of China's GHG Emissions

Chinese GHG emissions have grown rapidly. According to unofficial GHG estimates from China, from 1994 to 2004, China's annual average GHG growth rate was around 4%.[28] In this period, the share of carbon dioxide in total GHG emissions increased from 76% to 83%.[29]

The most recent estimates come from BP are somewhat higher than IEA's, globally and for China. According to BP's estimates for 2010, China's and the United States' CO_2 emissions were 8,333 and 6,145 MMT CO_2, respectively. BP estimates that China's CO_2 emissions grew 339% from 1990 to 2010 (Figure 3). (This compares with U.S. CO_2 emissions growth of 13% from 1990 to 2010.)

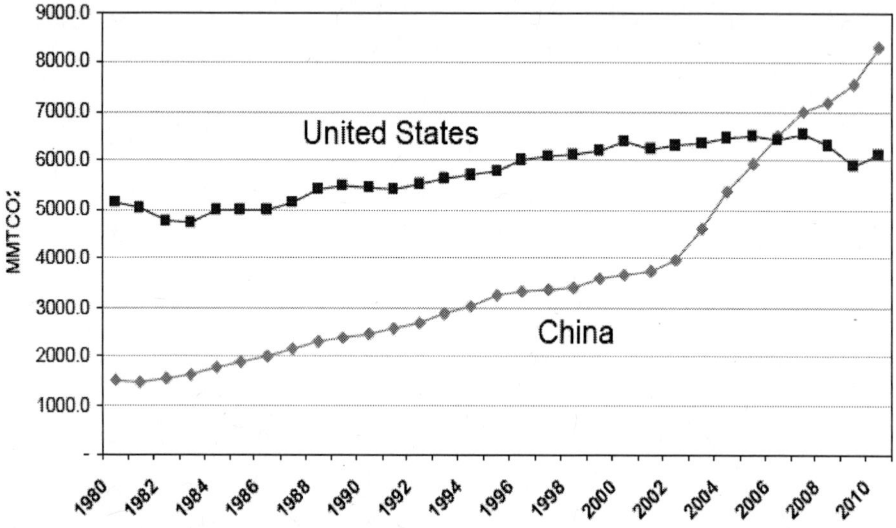

Source: BP, Statistical Review of World Energy 2011 at http://www.bp.com/sectiongenericarticle800.do?categoryId=9037188&contentId=7068766.

Figure 3. Growth of Estimated CO_2 Emissions from the United States and China (1980–2010).

GHG Emissions by Sector

Since China has not officially released a GHG inventory since 1994, it is difficult to know how sectors contribute to national emissions. The Pew Center on Global Climate Change estimated that, in 2003, electricity and heat made up 42% of China's GHG emissions,[30] industry[31] 21%, agriculture 20%, households and services 9%, transportation 5%, and waste 3%.[32]

According to IEA estimates, of China's 2005 GHG emissions, about 68% came from fuel combustion in all sectors (Figure 4). About 5% evaporated as methane from energy related systems. Another 10% came from industrial processes, and about 14% came from agriculture. Waste and miscellaneous sources accounted for the remaining 4% of China's GHG emissions that year.

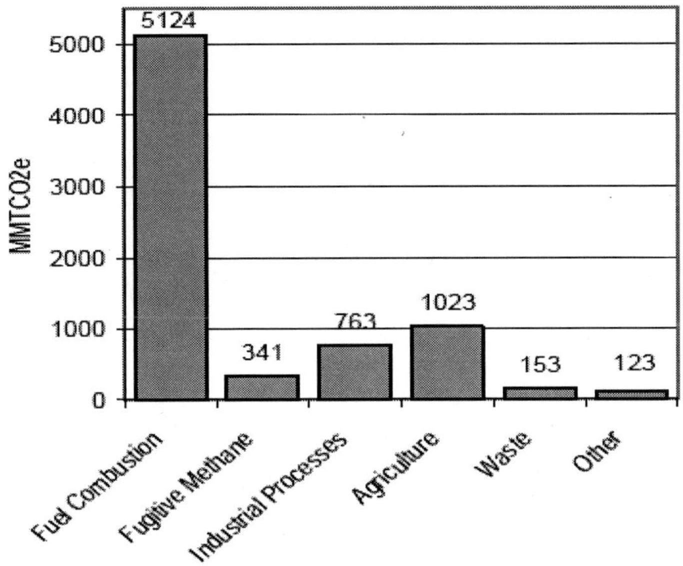

Source: CRS graph from IEA estimates, extracted July 8, 2008.

Figure 4. GHG Emissions By Source Types in China in 2005.

THE DRIVERS OF CHINA'S GHG EMISSIONS

China's GHG emissions are the highest in the world because of its very large economy, the high share of the economy generated by energy-intensive (and GHG emitting) industry, and the high share of coal in China's energy mix.

The Growing Economy

China's GHG emissions are the highest in the world primarily because of the size of its economy, which is, in turn, due in large part to its vast population of 1.3 billion people. (Per capita, production is much smaller than that of the United States. But multiplying China's low per capita production by its large population makes the China's economy is the second largest globally.)

China exercises strong policies to slow population growth; the growth of population in 2005 was approximately 0.6%, down from an average rate of about 1.1% in the 1990s and 1.4% in the 1980s.[33] The relatively slow rate of population growth helps to diminish the corresponding growth of national GHG emissions. China's population policies are clearly not aimed at mitigating GHG emissions, but observers note that without them emissions would have increased substantially.[34]

Though China's population has been growing relatively slowly, production per capita has grown rapidly. China's economy has experienced annual growth rates from 8 to 14% over the past decade. The Chinese economy is now the second largest economy in the world: int$9.1 trillion in 2009—two-thirds as large as the int$14.1 trillion U.S. economy (compared using Purchasing Power Parities), and more than twice the size of Japan's (the third largest economy).

Figure 5 provides one estimate of the economic activities that have driven recent growth of Chinese GHG emissions. It concludes that the emissions associated with exports have grown rapidly in the past few years, as well as capital investment (construction of buildings, roads, etc.) and increasing consumption by the population and their rising incomes.[35]

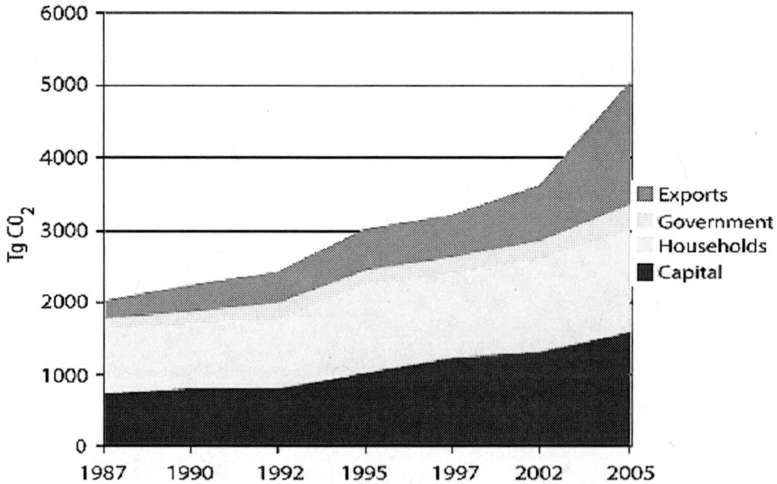

Source: Weber et al., op. cit.

Note: China's total domestic CO_2 emissions, divided by driving demands: exports, governmental consumption, household consumption, and capital investment.

Figure 5. One Estimate of Factors Driving Recent Growth of GHG Emissions in China.

China's Energy Sector

Energy production and use emit roughly three-quarters of all GHG in China, as in most other countries (Figure 4). China's rising GHG emissions have been primarily due to expansion of energy-intensive[36] industrial activity—largely in manufacturing for export and in construction of new infrastructure (Figure 5).[37] China's continuing heavy reliance on coal also results in high CO_2 intensity of its economy.

Energy Intensity

Chinese energy demand has surged since 1990, growing at a rate sometimes faster than the economy (Figure 6).[38] Years of government control over the energy sector, as well as incremental moves to decentralize the system, led to important distortions and inefficiencies. In the early 2000s, a decades-long reduction in energy intensity of the economy halted and reversed (Figure 7), increasing China's dependence on energy to fuel its economic growth. Simultaneously, energy shortages forced stoppages at some enterprises. Though demand for oil for electricity production peaked[39] then declined, demand by vehicle use soared, aggravating concerns about access to supplies from the Middle East and Russia, and about vulnerabilities to price fluctuations. By 2007, China became a net importer of coal. Also, choking levels of pollution, much from the energy sector, triggered public protests, while international pressure increased on China to control its related emissions of air pollutants and GHG. Though reforms were promoted from the 1990s, weak implementation, unreliable data and monitoring, and a predominant emphasis on economic expansion limited their effectiveness. These factors focused the attention of the central government on revising national energy policies in China's 11th and 12th (current) 5-year plans. (Policies will be discussed in "China's GHG Abatement Policies and Programs".)

Reliance on Coal in the Fuel Mix

China uses a high portion of coal in its fuel mix (Figure 8)—another reason that its CO_2 emissions are high relative to the size of its economy (its CO_2 intensity).

Coal emits far more CO_2 for the amount of energy it provides than other fossil fuels. Coal's "emission factor" is about 30% higher than that of crude oil, and about 70% more than natural gas, on average. So the high use of coal, and of fossil fuels more generally, in the Chinese economy explains why China's GHG emissions are proportionately high.

Compare the United States and China: In 2009, the United States' fuel mix comprised about 23% coal and 37% petroleum. Nuclear, hydroelectric, and other renewable sources contributed another 13%. In China, however, coal provided 70% of total energy, with petroleum contributing 18%, gas 4%, and nuclear, hydroelectric, and renewables providing the remaining 8%.[40,41] China now consumes about three times as much coal each year as the United States, even though its total energy demand in 2009 was about 6% higher, according to BP data.

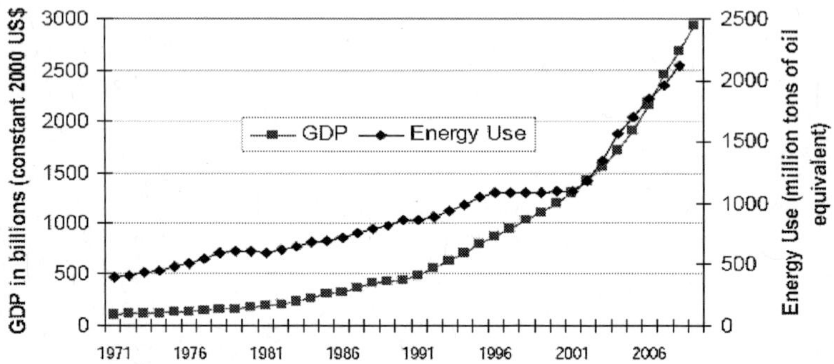

Source: CRS figure using data from World dataBank, *World Development Indicators*, extracted July 6, 2011.

Figure 6. Growth of the Economy and Energy Use in China (1971–2009).

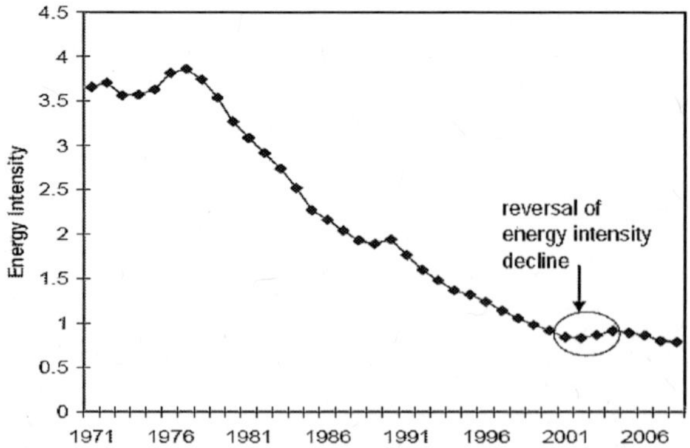

Source: CRS figure using data from World Development Indicators, extracted July 7, 2011.

Notes: Energy intensity measured as kilotons of oil equivalent per billions of constant 2000 US $.

Figure 7. China's Energy Intensity; (1971–2009).

The next biggest difference between China and the United States after China's greater reliance on coal is China's limited quantities of domestic natural gas, and its lower use of petroleum. (However, while oil is being backed out of electricity generation in China, demand for oil for transportation is surging with automobile use.) China also has many fewer nuclear power plants.

Although not reflected in China's CO_2 emissions estimates, the energy sector also emits a large portion of methane (CH_4) from coal production. Due to the dangers of uncontrolled methane leaks in mines, the government has tried to force capture and abatement of those methane emissions, but rates are thought to remain high (based, in part, on continuing frequency of coal mine explosions).

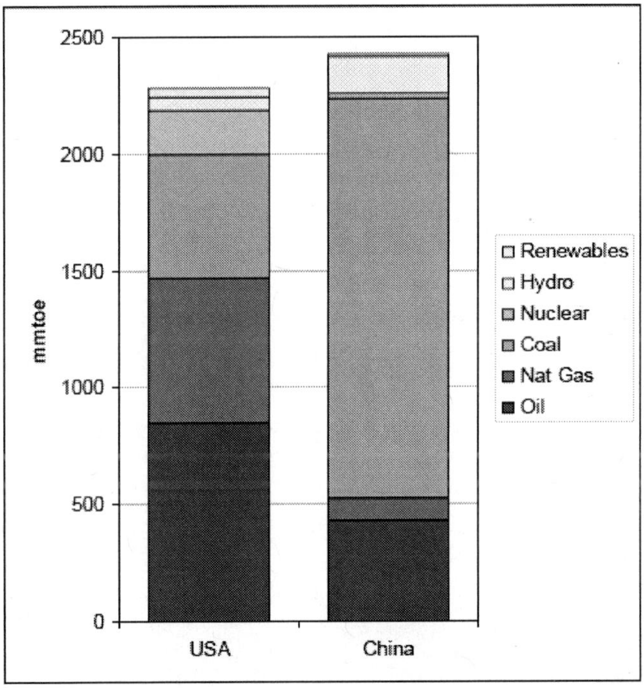

Source: CRS figured using data from BP, *Statistical Review of World Energy* (June 2011) at http://www.bp.com/sectionbodycopy.do?category Id=7500& contentId=7068481.

Figure 8. Primary Energy Consumption by Energy Types, United States and China; (2009).

Other Sectors Emitting GHG

Agriculture is the next most significant sector for GHG emissions, although recent estimates are not available. In 1994, China's agriculture sector contributed 17% of all GHG emissions, 50% of all methane emissions (especially due to rice cultivation, and pig and sheep production), and about 92% of estimated nitrous oxide emissions (due to fertilizer application).[42] It is likely that these percentages have shifted significantly in the intervening years, however.

China's GHG Abatement Policies and Programs

China has increasingly strengthened its policies and programs to curtail GHG emissions, culminating in a pledge under the 2009 Copenhagen Accord[43] to achieve a 40 to 45% reduction in its carbon intensity[44] by 2020. The government has sought to enforce and incentivize many programs to improve energy efficiency and expand the shares of non-emitting sources of energy.

In addition, it has promoted policies in agriculture and other sectors to abate GHG emissions. Still, some critics are skeptical of China's ability to achieve its targets, while others believe that China's efforts are little more than "business-as-usual." One recent analysis concluded that achieving the carbon intensity target is feasible, but would require strengthening of existing policies to stimulate energy efficiency in multiple sectors and to increase shares of renewable and nuclear electricity generation.[45] Under existing policies, most analysts expect China's GHG emissions to slow their rate of growth but to continue to increase until around 2030.

As discussed earlier, China's government has felt pressure to abate choking levels of local pollution as well as to engage with other large GHG emitters in international cooperation to reduce its emissions. China had policies for many years to improve efficiency, and even to tax polluting emissions, but enforcement and effectiveness of those policies were less than anticipated. By 2007, when China was under strong international pressure to negotiate GHG reduction targets under the UNFCCC for the period beyond 2012, China released its National Climate Change Program, a plan to address climate change.

The most challenging aspect of China's policy, arguably, was its goal to lower energy intensity 20% by 2010. By 2010, the government says the nation fell just short of that goal, with a 19.1% improvement. However, under severe scrutiny towards the end of 2010, it appears that many energy managers met their goals by stopping or slowing production rather than improving efficiency.[46] This seems to have resulted in a slight "rebound" of energy intensity early in 2011.

Chinese leadership has warned against such tactics to reduce energy intensity and tightened its objectives in the 12th 5-Year Plan, from 2011-2015. Besides the improvement in energy intensity, measures in the 11th 5-Year Plan resulted, according to China's claims, in

- The closure of thousands of small, inefficient, and polluting coal-fired power plants, iron and steel mills, cement kilns, aluminum plants, and others. China reports closures of old coal-fired power plants exceeding 71 GW of capacity in 2006-2010, and another 11 GW in the first half of 2011. The government says these plants would otherwise have emitted 164 $MMTCO_2$ annually.[47]
- The share of non-fossil energy reached 9.6% in 2010, up from about 7% in 2005. China became the largest wind power market by 2010, for both supply and use.
- The stock of carbon stored in forests increased by an estimated 13 billion cubic meters.[48]

The Chinese government has stated that some of the policies put in place to achieve China's targets for 2010 will likely continue into the future. Examples include

- investment in ultra-efficient coal fired electricity generation;
- improvement of existing coal-fired industrial boilers;
- closure of inefficient energy and industrial production capacity;
- expansion of combined heat-and-power;
- improvements of industrial motor efficiencies;
- standards for energy efficient lighting, buildings, and appliances;
- efficiency labels for appliances;
- improved enforcement of standards;
- financial incentives to build renewable energy capacity;
- requirements of feed-in tariffs to promote renewable energy generation;

- monetary awards for energy-saving achievements by companies and public institutions;
- vehicle efficiency standards that exceeded those of the United States;
- tightened efficiency standards for buildings and appliances, and forest coverage expanded to 20%.

Regarding CO_2 emissions and energy policies, the details of China's 12th 5-Year Plan (covering 2011 to 2015) and its implementing measures are still in draft. The national government has announced a number of its targets that, if achieved, would reduce growth of GHG emissions. By 2015:

- energy intensity should improve by 16% by 2015;
- carbon intensity should improve 17% by 2015, reaching 40-45% relative to 2005 levels by 2020;
- the share of non-fossil energy should reach 11.4% by 2015 and 15% by 2020;
- Forest coverage should increase by 12.5 million hectares (31 million acres) by 2015, and 40 million hectares (99 million acres) by 2020, compared to 2005 area;
- Nine pilot CO_2 cap and trade programs have been established across several cities;
- The length of high-speed railways is planned to increase to 45,000 km (27,962 miles).

Comprehensive laws to facilitate meeting those targets are now under consideration by the legislature. A variety of statements from the government indicate that some of the main policies China would use to achieve its goals are:

- continued economic restructuring toward higher value-added and less energy intensive production;
- financial awards to companies and public institutions for quantitative goals for saving energy and substituting alternative for fossil fuels in the transport sector;[49] and
- expansion of the "energy-saving service sector" with a focus on the transport and construction sectors, making energy efficiency a "criterion for market entry" and setting benchmarks for energy performance.[50]

Projected CO_2 Emissions

Just as China has not provided estimates of historical or current GHG emissions, it has not provided projections of future emissions. A variety of organizations have produced their own projections, with different assumptions about underlying "business as usual," efficacies of Chinese policies and programs, and rates of technological advance. In the longer term, Chinese officials are signaling possible absolute reductions in China's GHG emissions. As reported by the Xinhua news agency, a recent report by the Chinese Academy of Engineering concluded that

> China's energy development is projected to experience a "historic transition" around 2030 when its consumption of coal becomes restrained, the emission of carbon dioxide reaches its peak and energy-saving capacities around the world reaches an advanced level.[51]

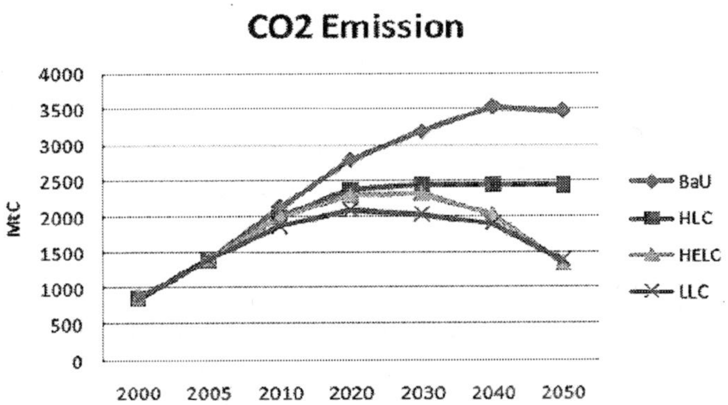

Source: Kejun et al. "Technology roadmap for low carbon society in China" *J. Renewable Sustainable Energy* 2, 031008 (2010).

Notes: The figure depicts four scenarios for China's CO_2 emissions: BaU represents "taking no climate change policies under the high GDP growth rate assumption." The HLC scenario represents unilateral Chinese actions to address energy security, domestic environmental standards, and a low carbon development strategy. The HELC scenario represents further technological efforts to abate CO_2 emissions, including large-scale deployment of Carbon Capture and Storage (CCS) in China. This scenario assumes China would act in the context of global efforts. All three of these scenarios assume high rates of economic growth in China. The LLC scenario assumes lower economic growth rates, along with efforts to transition to a low carbon economy. The reference paper provides more detailed assumptions.

Figure 9. One Projection of Future Chinese Emissions.

This conclusion is consistent with other analyses and official statements suggesting that 2030 could mark the high point of Chinese CO_2 emissions if current policies continue. One analysis that seemed linked to the Chinese pledge to improve carbon intensity by 40-45% by 2020 is illustrated in Figure 9. (A very similar figure was made available at the time China announced its pledge.)

Another recent analysis by Lawrence Berkeley National Laboratory (LBNL) concludes similarly that China's CO_2 emissions could begin to level off around 2030, due to saturation of some demand for greater energy services, as well as standards and incentives for more efficiency and non-fossil technologies.[52]

An alternative view is represented by the International Energy Agency's World Energy Outlook 2009. It suggests that China's current policies (before the 12th 5-Year Plan) would result in continually rising CO_2 emissions, reaching 12.6 $MMTCO_2$ by 2035.[53] That would be almost a doubling of China's 2005 emissions.

China's Stance on International Obligations

China and other countries that had low incomes in 1992 were exempted in the United Nations Framework Convention on Climate Change (UNFCCC) and the 1997 Kyoto Protocol from taking on quantified GHG reduction obligations, based on the principle of "common but differentiated responsibilities" contained in the Convention. While Parties agreed that the already industrialized Parties ("Annex I Parties") should take the first steps in abating their GHG emissions, it has been clear from a scientific standpoint that the objective of the UNFCCC—to stabilize atmospheric concentrations of GHG at a level that would avoid dangerous anthropogenic interference with the climate system (in Article 2)—could be met only when all significant emitters reduce their net emissions (i.e., emissions minus removals by photosynthesis) to near zero. China and other low income economies, however, resisted any discussion of when and how they might take on GHG obligations. At the same time, the United States and a few other countries rejected taking on GHG mitigation commitments unless *all* major emitters take on commitments. Impasse continued until agreement was reached on the 2009 Copenhagen Accord: countries associated with the Accord submitted pledges to GHG targets and mitigation actions they would take. These pledges are politically but not legally binding. Agreement on legally binding commitments under the UNFCCC or the Kyoto Protocol seems unlikely for the foreseeable future.

China's stance against legally binding obligations rests on several points. First, China argues that the existing, elevated concentrations of GHG in the atmosphere are due to historical emissions from the already industrialized countries, such as the United States. (The "Brazil Proposal" of 1997 suggests that developing countries should not take on GHG abatement requirements until their accumulated contributions to atmospheric concentrations equally the contributions of the Annex I Parties—in other words, not for decades.) China's large and rapidly growing emissions, however, may soon places its contributions to atmospheric concentrations on a par with the historic contributions of many smaller Annex I Parties. Also, China has extended its position on differentiation of obligations to cover GHG emission reporting and review, not just abatement responsibilities. As discussed earlier in the section on China's GHG emissions, that country's reticence to be transparent about its emissions and sequestration and quantification of its policies and programs, has been an important point of conflict in the UNFCCC negotiations.

Second, China and other lower income countries point out that the industrialized economies benefited from essentially unconstrained use of energy to fuel their growth; lower income countries should also be exempt from constraints on their use of energy until their incomes have caught up. They also contend that, if global carbon emissions must be limited, each person should have an equal "right" to emissions; this would imply that the "rights" of each American would be small fraction of their current actual emissions, while citizens of developing would be allowed to expand their average emissions. (Even under this line of argument, however, China would need to reduce its per capita GHG emissions in scenarios that would stabilize GHG concentrations at many moderate, proposed targets.) An "equity" proposal rumored to be forthcoming from India, China, Brazil and South Africa would set a principle in the negotiations that a country would not take on GHG abatement obligations until each of its citizens had access to energy and had emerged from poverty. Such a proposal could illuminate the widely varying conceptions of "equity" among Parties and individuals, as well as perhaps being inconsistent with achieving the UNFCCC's objective of stabilizing GHG concentrations.

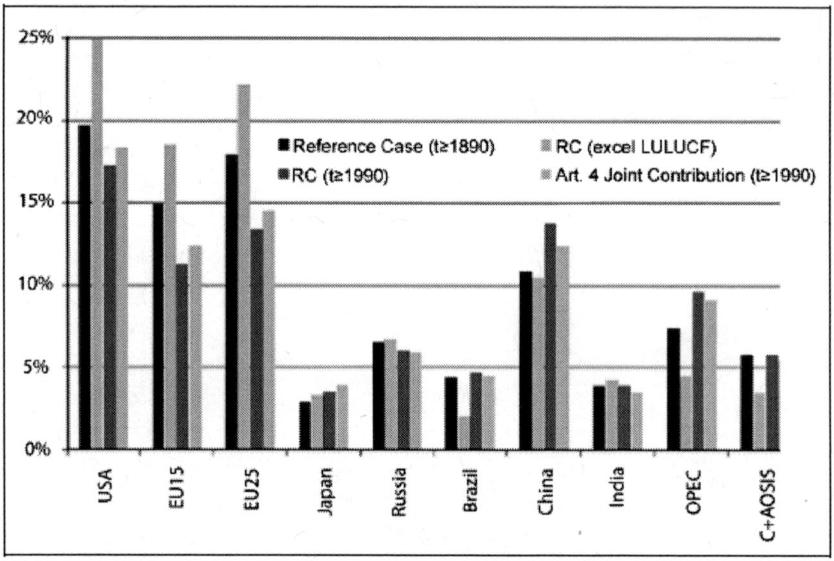

Source: Muller et al. “Differentiating (historic) responsibilities for climate change” *Climate Policy*, Volume 9, Number 6, 2009 , pp. 593-611(19).

Notes: This figure was derived from analyses commissioned by the UNFCCC Parties, as Modelling and Assessment of Contributions to Climate Change (MATCH). Reference Case (t> 1890) calculates the contributions beginning in the year 1890, while RC (t> 1990) makes the calculation with emissions beginning in 1990. RC (excel LULUCF) excludes emissions from land use, land use change and forestry (LULUCF).

Figure 10. Relative Contributions to Climate Change in 2000 Under Alternative Assumptions.

In addition, China and other non-Annex I Parties have underscored the greater financial capabilities of the wealthier countries to undertake GHG abatement, while China’s priority must be to alleviate poverty and to raise average incomes towards those of the Annex I Parties. China, for decades, has sought financial and technological assistance as part of nearly every internati issue. China’s robust economy, large foreign reserves, and leading experience in manufacturing and deploying many advanced technologies has reduced the credibility of its requests. In contrast, China has increased its engagement in technology cooperation in recent years, including with the United States.

For many years, China and other non-Annex I Parties allied to block discussion of new commitments for non-Annex I Parties. By 2009 and the Copenhagen negotiations, however, some countries that feel vulnerable to the impacts of climate change began to perceive that it would not be in their interests to sustain the "no new commitments for developing countries" mantra. The fractious negotiations in Copenhagen spotlighted these diverging interests and added to pressure for China and others to agree to the political pledging processes that have emerged. For the foreseeable future, however, China is likely to continue its opposition to taking on legally binding GHG targets, as well as to enhanced requirements for GHG reporting and international review of its policies and progress.

End Notes

[1] For more information on international negotiations on climate change, see CRS Report R40001, *A U.S.-Centric Chronology of the International Climate Change Negotiations*, by Jane A. Leggett.

[2] The industrialized countries in 1990 included the Members of the Organization for Economic Cooperation and Development (OECD: the United States, Canada, Japan, Australia, New Zealand, the European Union, the Nordic Countries, and Turkey), the former Soviet Union, and Eastern European countries.

[3] World Resources Institute, *Climate Analysis Indicators Tool (CAIT)*, at http://cait.wri.org/. Data extracted July 11, 2011.

[4] Carbon dioxide (CO_2) is, by far, the most important human-related GHG (though water vapor, which is believed not to be directly related to human activities is the most important GHG in the atmosphere), and (CO_2) emissions from fossil fuel-related production, transmission, storage, and use represents in 2010 about three-quarters of all human-related GHG emissions. Estimates of most other human-related GHG are less available and less reliable than estimates for CO_2 emissions. This report will report total human-related GHG emissions when possible, but frequently will use CO_2 emissions when data for all GHG are not reliable (in CRS' judgment).

[5] Ultimately, stabilizing atmospheric GHG concentrations means stopping their accumulation in the atmosphere. This requires, essentially, that net GHG emissions must fall to about zero. Although some countries might achieve "negative emissions" by enhancing removals (e.g., growing more forests), this could not compensate for growing emissions in other countries. All countries on average would have to bring their net emissions down to net zero. A recent example of related analysis is found in H. David Matthews and Ken Caldeira, "Stabilizing climate requires near-zero emissions" *Geophysical Research Letters* Vol. 35, L04705, doi:10.1029/2007GL032388 (2008) at http://www.precaution.org/lib/zero_emissions_required.080227.pdf.

[6] Lin Erda; Xu Yinlong; Wu Shaohong; Ju Hui; Ma Shiming, "Synopsis of China National Climate Change Assessment Report (II): Climate Change Impacts and Adaptation." *Advances in Climate Change Research*. (February 2006). original at http://en.cnki.com.cn/Article_en/CJFDTOTAL-QHBH200602001.htm.

[7] See, for example, Joanna I. Lewis, "Climate change and security: examining China's challenges in a warming world" *International Affairs* 85: 6 (2009) 1195–1213 at http://www.chathamhouse.org.uk/files/15111_85_6lewis.pdf.

[8] See CRS Report RL33534, *China's Economic Conditions*, by Wayne M. Morrison.

[9] Using data from the International Monetary Fund at IMF.

[10] See CRS Report RL34314, *China's Holdings of U.S. Securities: Implications for the U.S. Economy*, by Wayne M. Morrison and Marc Labonte.

[11] S. Chen and M. Ravallion, "China is Poorer Than We Thought, but No Less Successful in the Fight Against Poverty," Policy Research Working Paper, WPS-4621, World Bank, May 2008, Table 2.

[12] Purchasing Power Parities (PPPs) compare financial data across countries using indices of relative costs rather than market currency exchange rates. "International dollars" are a hypothetical currency that uses PPPs equalizing the US$ and the international $ (int$) at a specific point in time.

[13] World Bank, World dataBank, data extracted July 11, 2011, at http://databank.worldbank.org/ddp/home.do?Step=1& id=4.

[14] See footnote 12.

[15] See CRS Report RL33416, *Social Unrest in China*, by Thomas Lum.

[16] World Bank. 2007. *Cost of Pollution in China: Economic Estimates of Physical Damages*. Washington DC.

[17] By a "recycling" economy, the Chinese mean that they would move away from an economic system of resource extraction, processing, and disposal of materials as waste, toward reuse and recycling of materials already existent in the economy.

[18] Garrett Hardin, "The Tragedy of the Commons," *Science* 162:1243-1248, 1968.

[19] There are many ways in which efforts may be divided up internationally, through market systems that set a price on "use" of the resource (i.e., emitting pollution being equivalent to using the atmosphere for waste disposal) or by agreement to emission quotas or "caps" or other means.

[20] The *National Climate Change Program*, released by the Chinese government in 2007, identifies many potential impacts of climate change on China. Additional examples of recent publications outlining such costs include A. Thomson, R. Izaurralde, N. Rosenberg, and X. He, "Climate Change Impacts on Agriculture and Soil Carbon Sequestration Potential in the Huang-Hai Plain of China," *Agriculture Ecosystems & Environment* 114 (2-4): 195-209, 2006; X. Wang, F. Chen, and Z. Dong, "The Relative Role of Climatic and Human Factors in Desertification in Semiarid China," *Global Environmental Change—Human and Policy Dimensions* 16 (1): 48-57, 2006; X. Zhang, and W. Liu, "Simulating Potential Response of Hydrology, Soil Erosion, and Crop Productivity to Climate Change in Changwu Tablel and Region on the Loess Plateau of China," *Agricultural and Forest Meteorology* 131 (3-4): 127-142, 2005.

[21] *The People's Republic of China Initial National Communications on Climate Change*, Office of National Coordination Committee on Climate Change, 2007, available at http://www.ccchina.gov.cn/en/index.asp.

[22] Like some other developing countries, China has resisted proposals in international negotiations under the United Nations Framework Convention on Climate Change (UNFCCC) that developing countries submit annual GHG inventories to the Conference of the Parties.

[23] "China: Backgrounder: Current GHG Emissions in China, Xinhua, June 4, 2007, http://www.chinaview.cn.

[24] CO_2e means "carbon dioxide equivalent," which is an aggregate of all greenhouse gases with each gas weighted by its effect on temperature change compared to CO_2.

[25] China published this document in June 2007 to outline its plan to address climate change and show its determination to mitigate emissions. The official document is available at http://english.gov.cn/2007-06/04/content_635624.htm.

[26] National Development and Reform Commission (NDRC), "China's National Climate Change Program," p. 10. Note, afforestation is the establishment of a forest on lands not previously forested, as opposed reforestation.

[27] BP, op. cit.

[28] The Climate Change Group, "China Briefing," Issue 1, September 2007.

[29] Xinhua, op. cit., June 4, 2007.

[30] Most the electricity demand in China is for its industrial sector.

[31] Not associated with electricity and heat.

[32] Pew Center on Global Climate Change, "Climate Change Mitigation Measures in the People's Republic of China," p. 1, April 2007.

[33] Population growth rates taken from United Nations, *World Population Prospects: The 2006 Revision*, 2007; and Central Intelligence Agency, *The World Factbook: China*, July 2008, https://www.cia.gov/library/publications/theworld-factbook/geos/ch.html.

[34] According to Gao Guangsheng of the NDRC, Chinese carbon dioxide emissions would have been 1.3 billion tons higher in 2005 had the country not implemented family planning policies in the 1970s, http://www.pewclimate.org/docUploads/Gao%20Guangsheng.pdf. Some people find family planning and other methods of birth control immoral. Regardless of moral viewpoints on China's family planning policies, slowing population growth is likely to contribute to GHG mitigation.

[35] Weber, Christopher L., Glen P. Peters, Dabo Guan, and Klaus Hubacek. 2008. "The Contribution of Chinese Exports to Climate Change." *Energy Policy*. doi:10.1016/j.enpol.2008.06.009.

[36] Energy intensity is the amount of energy consumed per unit of production in an economy. In Table 1, energy intensity is measured as thousands of tons of oil-equivalent used per unit of Gross Domestic Product (GDP). As economies develop, they frequently shift from an agricultural base to resource extraction and manufacturing, which increases their energy intensities. With further growth, the shares of higher value-added production and services grow, resulting typically in a decline of energy intensity.

[37] D. Rosen and T. Houser, op. cit.

[38] For a discussion of China's surging energy use, see D. Rosen and T. Houser, *China Energy: A Guide for the Perplexed*, China Balance Sheet, May 2007, http://www.iie.com/publications/papers/rosen0507.pdf

[39] With shortages in electrical power from large installations, demand for diesel fuel for small scale electricity generation soared in the early 2000s, peaking in 2004. Rapid construction of electric power plants along with other government policies has alleviated that demand for oil for electricity and its share has fallen to less than 1%.

[40] These data are from the *BP Statistical Review of World Energy 2011*, available at http://www.bp.com.

[41] Estimates of GHG emissions in China are uncertain, in part because of underlying uncertainties in official energy and economic data. Such data issues are more pronounced when considering data over time (for example, regarding changes in coal data over the past decade) or in comparison to other countries, where levels of uncertainty may be less.

[42] Dong Hongmin "China GHG emission from agricultural activity and its mitigation strategy" (2007) at http://www.globalmethane.org/expo_china07/docs/postexpo/ag_hongmin.pdf.

[43] See CRS Report R40001, *A U.S.-Centric Chronology of the International Climate Change Negotiations*, by Jane A. Leggett for more information.

[44] Carbon intensity is the quantity of annual emissions relative to annual economic production, typically measured as Gross National Product.

[45] Nan Zhou, David Fridley, Michael McNeil, Nina Zheng, Jing Ke, and Mark Levine "China's Energy and Carbon Emissions Outlook to 2050," LBNL Report LBNL-4472E (April 2011).

[46] http://www.ccchina.gov.cn/en/NewsInfo.asp?NewsId=27352.

[47] Industrial Fuels and Power, "China succeeds in meeting target for obsolete coal-fired power plant closures, coal consumption falling back on abundant supply" (July 26, 2011) at http://www.ifandp.com/article/006210.html.

[48] http://www.f-paper.com/?i469938-Photo:-Sun-Cuihua:-Chinas-greenhouse-gas-control-policies- and-marketmechanisms

[49] http://www.ccchina.gov.cn/en/NewsInfo.asp?NewsId=28763.

[50] http://www.ccchina.gov.cn/en/NewsInfo.asp?NewsId=28694.

[51] Chinese Academy of Engineering, "Research Report on China's Mid-term and Long-term Energy Development Strategy" (2011) reported at http://news.xinhuanet.com/english2010/china/2011-02/28/c_13754385.htm.

[52] Nan Zhou, David Fridley, Michael McNeil, Nina Zheng, Jing Ke, and Mark Levine "China's Energy and Carbon Emissions Outlook to 2050," LBNL Report LBNL-4472E (April 2011). See Figure ES-2, for example.

[53] Slide 30 at http://www.energy.eu/publications/weo_2010-China.pdf.

In: Going Green in China: Policy and Potential ISBN: 978-1- 62100-691-6
Editors: Spencer B. Talscott

Chapter 2

A LOW CARBON DEVELOPMENT GUIDE FOR LOCAL GOVERNMENT ACTIONS IN CHINA*

Nina Zheng, Nan Zhou,
Lynn Price and Stephanie Ohshita

DISCLAIMER

This document was prepared as an account of work sponsored by the United States Government. While this document is believed to contain correct information, neither the United States Government nor any agency thereof, nor The Regents of the University of California, nor any of their employees, makes any warranty, express or implied, or assumes any legal responsibility for the accuracy, completeness, or usefulness of any information, apparatus, product, or process disclosed, or represents that its use would not infringe privately owned rights. Reference herein to any specific commercial product, process, or service by its trade name, trademark, manufacturer, or otherwise, does not necessarily constitute or imply its endorsement, recommendation, or favoring by the United States government or any agency thereof, or The Regents of the University of California. The views and opinions of authors expressed herein do not necessarily state or reflect those of the United States

* This is an edited, reformatted and augmented version of Ernest Orlando Lawrence Berkeley National Laboratory Publication, LBNL-4839E, dated June 2011.

ABSTRACT

Local level actions are crucial for achieving energy-saving and greenhouse gas emission reduction targets. Yet it is challenging to implement new policies and actions due to a lack of information, funding, and capacity. This is particularly the case in developing countries such as China. Even though national energy intensity and carbon intensity targets have been set, most local governments do not have the knowledge regarding actions to achieve the targets, the cost-effectiveness of policies, the possible impact of policies, or how to design and implement a climate action plan.

This paper describes a guidebook that was developed to motivate and provide local governments in China with information to create an action plan to tackle climate change and increase energy efficiency. It provides a simple step-by-step description of how action plans can be established and essential elements to be included—from preparing a GHG emission inventory to implementation of the plan. The guidebook also provides a comprehensive list of successful policies and best practices found internationally and in China to encourage low carbon development in industry, buildings, transportation, electric power generation, agriculture and forestry.

This paper also presents indicators that can be used to define low-carbon development, as well as to evaluate the effectiveness of actions taken at an aggregated (city) level, and at a sectoral or end use level. The guidebook can also be used for low carbon development by local governments in other developing countries.

INTRODUCTION

Local level actions and leadership are crucial for achieving national energy-saving and greenhouse gas emission (GHG) reduction targets. Local level actions can also assist in proving the effectiveness of new policies or initiatives by demonstrating them at a smaller scale. It is often also shown that innovative policies or practices can be relatively easily implemented at the local level because of the reduced scale and the possibility of exemption from national legislative bureaucracy. Following success at the local level, the pilot policies or practices can be replicated in other localities or expanded to a national program. For example, China's Top-1000 Enterprise Program drew upon the successful experience from a demonstration program implemented in two steel mills in Shandong province that was modeled after the voluntary agreement programs in The Netherlands and the UK (Price et al., 2003).

In developed countries, state and local level initiatives have proven to be very successful in transforming markets by engaging businesses and educating citizens. In the U.S., many states, cities and counties have forged ahead with dedicated funding and strategic policies to promote energy efficiency and renewable energy. California -- one of the best examples -- has set ambitious energy efficiency and GHG emission reduction targets and has implemented stringent, innovative policies and actions to improve energy efficiency and reduce emissions. As a result, California's per capita electricity consumption is 40% less than the average per capita electricity use in the U.S. California's experiences have often been replicated or echoed in other states, and some of the policies eventually became national regulations. Many federal appliance standards today are the direct result of such state leadership (REEEP, 2010). In addition, the experience also demonstrates that the adoption of a comprehensive energy and climate plan can stimulate the local economy, create green jobs, as well as produce new revenue (Roland-Holst, 2008).

Even so, it is challenging to initiate and implement new policies and actions due to a lack of information, funding, and capacity. This is particularly the case in China. Even though national energy intensity and carbon intensity targets have been set, most local governments do not have the knowledge regarding what they can do to achieve these goals, the cost-effectiveness of policies, the possible impact of policies, or how to design and implement a climate action plan.

Regarding low carbon development in China, the Chinese government announced a policy (Notice of Piloting Low-Carbon Provinces and Low-Carbon Cities) for establishment of Low-Carbon Cities and selected five provinces and eight cities as pilots in August 2010. The five provinces are Guangdong, Liaoning, Hubei, Shaanxi and Yunnan; and the eight cities are Chongqing, Shenzhen, Xiamen, Hangzhou, Nanchang, Guiyang, and Baoding. The policy outlines the following activities[1]:

1) Develop Low-Carbon Development Plan
2) Establish supporting policies to support low-carbon development
3) Establish low-emission industries

4) Establish GHG data collection and management systems
5) Promote low-carbon/green lifestyle and consumption model.

As more attention is being paid to low-carbon cities and in response to the goal of reducing carbon intensity by 40-45% by 2020 from the 2005 level, many other cities or counties are also following the trend toward low carbon development. However, many of these so-called "low carbon cities are actually "high carbon" as no consistent definition, or scientific scope or indicators have been developed. Some supposedly low-carbon cities built wide roads; although lined with beautiful trees, the roads encouraged *more* vehicle use. Some cities excluded imported electricity from their carbon accounting. Thus, it is important to clearly define indicators, standardize the development process, and indentify policies, programs, technologies and measures that can be undertaken to realize carbon emission reductions (or carbon intensity reductions) in participating cities.

This paper describes a guidebook that was developed to motivate and provide local governments in China with information to create an action plan to tackle climate change and increase energy efficiency. It provides a simple step-by-step description of how action plans can be established and essential elements to be included—from preparing a GHG emission inventory to implementation of the plan. The guidebook also provides a comprehensive list of successful policies and best practices internationally and in China to encourage low carbon development in industry, buildings, transportation, electric power generation, agriculture and forestry. Where available, the GHG emission reduction potential and cost-effectiveness of policies are also provided.

This paper also presents indicators that can be used to define low-carbon development, as well as to evaluate the effectiveness of actions taken at an aggregated (city) level, and at a sectoral or end use level. The paper does not intend to provide independent evaluation or analysis of the GHG emission reduction or cost-effectiveness of policies, but rather to synthesize and organize policy options based on existing literature, documents and reports. The guidebook can also be used for low carbon development by local governments in other developing countries.

LOW-CARBON INDICATORS

This study focuses on carbon, and defines a low-carbon city as a city that is actively and significantly lowering carbon emissions, even as its economy is maturing. Therefore the report gives guidance on the reduction of the two main carbon-based GHGs: carbon dioxide (CO_2) and methane (CH4).[2] CO_2 emissions are primarily from energy consumption of fossil fuels, as well as from non-energy industrial processes (like cement production), and forest loss. Methane emissions arise from agriculture (especially rice production), animal husbandry, other land use, industry (e.g., coal-bed methane), and waste decomposition.

Indicators are used to define a low-carbon city, to help cities explore the gaps and potential for carbon emissions reduction, to evaluate progress in implementing low-carbon development actions, and to compare or benchmark across cities. Key indicators identified in the guidance document are shown in the table below.

Sectors	Key Indicators
Aggregated and mixed indicators	energy or carbon intensity of the economy, i.e., overall energy or CO_2 per unit GDP, city greenhouse gas emission inventory
Aggregated relative indicators	energy or CO_2 per capita or per land area
Structural indicators	sectoral shares of energy and GDP
Residential and commercial sector	energy or CO_2 per floor area or per person/employee, percent compliance with building efficiency codes, Registered and certified LEED buildings, green buildings or other certification/capita, Installed capacity of integrated renewable or CHP in buildings/m2, space heating intensities
Industrial sector	physical efficiency (energy or carbon per ton of product); economic intensity (energy or carbon per unit value added)
Power sector indicators	CO_2 per kWh share of renewables in electricity supply
Transportation	primary energy or CO_2 per person-kilometer or ton-kilometer traveled; urban density; public transit use; kilometers of public transit per 100,000 population
Land Use and Waste Management	area share of mixed–use zoning (residential and commercial); percentage of landfill gas (methane) that is captured area share of green space and agricultural land; waste generated per capita; and recycling rate of waste.
Economic and Social Indicators	green jobs; income distribution and income per capita; and housing affordability.

Below are detailed description of the key indicators:

Aggregated

- Energy or CO_2/GDP
  The ratio of energy consumption, or carbon emissions, to gross domestic product (GDP) is used to measure the economic energy intensity of national economies. This is a key indicator used in China's 11th and 12th Five-Year Plans and announced internationally. An intensity indicator is appealing at the aggregate level, as it utilizes sets of data already tracked: energy, carbon, and GDP. However, this is a mixed indicator, accounting for both physical energy efficiency and economic structure that influences energy consumption. As economic development proceeds, the economic energy intensity typically declines, yet absolute energy and carbon still increase. Thus, there are significant limitations to this indicator and its use in target setting. Indicators distinctly focused on physical energy and carbon intensity, and on aspects of economic structure that affect energy consumption and carbon emissions, are encouraged instead.
- Energy or CO_2/capita
  Because energy consumption and carbon emissions can be strongly influenced by the size of the population, per capita indicators provide a better and more equitable basis for comparison across cities, provinces, and countries. This indicator is widely used in China and internationally. Highly aggregated per capita indicators, such as total energy or CO_2 per person, should be used with caution, however. A city with heavy industry and small population, which supplies other cities with cement and steel, would have high energy per capita. Yet the people of the city might use relatively little energy in their residences. Thus it is important to consider residential energy per capita, and the energy structure of a city, as well as total energy or CO_2 per capita.
- Energy or CO_2/land area
  Another measure of the energy or carbon intensity of a city can be a spatial measure, such as density per land area. This indicator is less common, but is being examined as cities consider how density of development influences energy consumption and carbon emissions.

- Economic Structure: sectoral shares of GDP (primary, secondary, tertiary)
  Because different sectors of the economy have notably different energy and carbon intensity, economic structure is an important indicator of structural influences on consumption and emissions. Of many definitions of economic structure, the simplest and most often used in China is the share of primary, secondary, and tertiary sectors of the economy. The secondary sector represents industry and construction – the most energy intensive—while the tertiary sector represents commerce and service-focused businesses such as Information Technology (IT), communication services, health care, and energy saving services. Even this fairly aggregated indicator can help cities identify areas for low-carbon development.
- Energy Structure: sectoral shares of energy consumption industrial, residential, transport, other Similar to an economic structure indicator, energy structure helps to identify areas needing extra attention for low-carbon development. Typical definitions of energy sectors include: industrial, residential, transport, agriculture and forestry, commercial, construction, etc. The first three sectors are the easiest to obtain data on; often the remaining energy sectors are grouped into "other energy." The industrial energy sector coincides with the secondary economic sector, while the others have overlap to different extents.
- Overall renewable energy use, and share of renewable in total city energy A key indicator of the carbon intensity of a city is the amount of renewable energy in the fuel mix. Renewable energy includes wind, solar thermal (for hot water), solar photovoltaic (for electricity), hydropower (with a distinction between large and small hydro), geothermal, and biomass. (Nuclear energy, although low carbon in operation, is not a renewable energy source.) This overall renewable energy indicator captures renewable that provide heat and transportation, as well as electricity. It is a well-defined indicator and helps a city clearly track progress toward low-carbon energy supply.

Buildings

- *Residential energy consumption or CO_2 emissions/capita (could be climate adjusted)* This indicator is commonly used to capture the energy efficiency of the building and home appliances, as well as behavior. If possible, it should be climate adjusted to exclude the influence on energy use induced by different climates. If climate adjusted, this indicator can be used for comparison with other peer cities. Weather variation can be accounted for by calculating cooling degree-hours (CDH) and heating degree-days (HDD).

- *Residential energy consumption or CO_2 emissions/m^2 (could be climate adjusted)* This indicator is similar to the per capita based intensity indicator, but targets building energy efficiency based on households without considering the number of people per household. Floor space data are available in the statistics in the form of m^2 per capita in urban and rural areas.
- *Commercial energy consumption or CO_2 emissions/m^2 (could be climate adjusted)* This indicator is commonly used for evaluating low energy and low carbon buildings. As with residential buildings, climate-adjusted values would more accurately reflect energy efficiency, taking into account the various heating and cooling demand among cities. Commercial and industrial floor space data may be collected through the local taxation office through property taxes. If data broken out by building types are available, then more detailed information and comparisons can be provided.
- *Energy/ electricity consumption per employee* In addition to the intensity in terms of energy use per unit of service floor area stated above (Commercial energy consumption or CO_2 emissions/m^2), an alternative indicator is energy use per employee in the sector. Data on employee numbers may be easier to collect than m^2, providing an advantage for the use of this indicator. However, energy or electricity consumption per square meter is slightly more meaningful as energy use is generally based on floor space.
- *Registered and certified LEED buildings, green buildings or other certification/capita* The goal of LEED or green buildings is to increase the efficiency with which buildings use resources, energy, water and materials. More such certified buildings in a city would imply lower energy and lower emission. However, the selection of the certification/label scheme should be evaluated as in practice, a well certified building could use more energy compared to a similar building without it. Research also shows that the measured energy performance of LEED buildings has little correlation with certification level of the building, or the number of energy credits

achieved by the building at the design phase (Newsham et al., 2009). Therefore the types of green certificates should be carefully assessed and selection should be based on verified actual savings that a certified building could deliver through measurement or tracking the change in utility bills.

- *Installed capacity of integrated renewable in buildings/m^2 (PV, solar water heaters, geothermal)* This indicator could demonstrate the cities' actions in reducing carbon emissions in buildings, not necessarily improving the efficiency of the buildings. Therefore, this indicator should be used in conjunction with other building efficiency measures that aim to reduce demand. By using the per m2 basis, the indicator can avoid favoring large buildings and large cities.
- *Installed capacity of combined heat and power (CHP) in buildings*
  On-site generation with recovery of waste heat can improve the energy efficiency in electricity use in buildings from 40% to 80%, and significantly reduce the carbon emissions. Particularly it will be very effective to the fast growing city center where the expansion of existing centralized grid reaches its limit due to the already congested infrastructure. The installed capacity of CHP in buildings could help to define the building energy performance.

Industry

- Share of coal in industrial fuel mix
  This indicator measures the extent to which lower carbon fuels, such as natural gas or biomass, are used by industrial facilities. Some industries, like cement, can partially replace the use of coal in the kiln with natural gas and many other alternative fuels including agricultural and non-agricultural biomass (sewage sludge, paper sludge, waste paper).
- Industrial economic energy or carbon intensity
  This indicator is often used at a highly aggregated level, combining all industrial energy consumption (and carbon emissions) with value-added economic output from industrial activities. This indicator can also be used at a sub-sectoral level, for example, to compare the intensity of overall cement production in a city with the intensity of chemicals or steel production or other industrial sub-sectors. This can

help a city identify development to promote or restrict, and help to benchmark with other cities.

- Enterprise-level resource productivity
- Manufacturing energy use per unit of manufacturing value added GDP
    - Sales/CO_2 emissions
    - Production volume/CO_2 emissions

  Productivity efficiency ratios measure the outputs of an enterprise in relation to their carbon emissions impacts. Sales/CO_2 emissions measure resource productivity, and production volume/CO_2 emissions measures process eco-efficiency.
- Physical energy or carbon intensity of key products produced
  Physical energy or carbon intensity (vs. economic energy or carbon intensity using value added as the denominator) such as energy use or carbon emissions/t crude steel or cement is preferred because it is a more robust indicator of the level of energy or carbon efficiency of the production process.
- CO_2 emissions per unit of manufacturing energy use and CO_2 emissions per unit of manufacturing GDP
  Since the manufacturing sector is extremely diverse, the International Energy Agency (IEA) relies more heavily on economic indicators such as energy use per unit value added, to be able to compare trends across manufacturing sub-sectors.

Electric Power

- CO_2 emissions per unit of generated electricity
  The amount of CO_2 equivalent emissions per unit of generated electricity is a common indicator for tracking de-carbonization of electricity supply. Expressed as tCO_2eq/kWh or tCO_2eq/TWh, this indicator helps to track the reduction of the use of carbon-intensive coal, and the fostering of renewable power generation as well as generation from natural gas and nuclear. This indicator also serves as an emission factor for determining carbon emissions from electricity use.
- Share of renewable sources in electricity supply; GW of operating renewable capacity The share of renewable sources in electricity supply is an indicator often used to track progress with Renewable

Portfolio Standards (RPS). While the CO_2eq/TWh is an indicator of net carbon in electricity, the share of renewable energy highlights the most sustainable energy sources. As the absolute amount of energy consumption changes, it is also useful to track the absolute amount of renewable generation capacity that is installed and actually operating. Wind farms cannot reduce carbon emissions if they are not sending power to the grid, so it is important to know and count the operational status of renewable energy sources.

Land Use and Waste Management

- Waste per capita (disposed, diverted)
The amount of waste per person is an indicator of the resourcefulness of the city's people; less waste means more efficient use of resources. Waste contains embodied energy and carbon, materials, and pollutants. By reducing the amount of waste generated—and disposed—per person, a city can save resources. Reducing the amount of waste going to landfill also reduces emissions of methane (a potent GHG) from the decomposition of organic waste in landfills.
- Recycling, and overall diversion rate of waste away from landfills
For any waste that is generated, there is still the possibility of recycling, composting, or otherwise diverting the waste from disposal. For most wastes, there is a net energy and carbon saving from recovering the waste. Tracking the recycling and/or diversion rate is a widely used indicator at the city and enterprise level.
- Percentage of landfill gas (methane) that is captured
For any waste that does end up in a landfill, there is one last option for recovering energy and reducing carbon emissions: landfill gas capture. Landfill gas is primarily methane; thus it can be captured to avoid methane emissions, and also utilized as a fuel, thereby replacing the use of a more carbon-intensive fuel such as coal.
- Agricultural Land - Hectares of food production, and share of agriculture in total area The amount of agricultural production within city (or county) boundaries is important to low-carbon development for two main reasons. The first is that local food production can lower energy and carbon, by reducing transport and refrigeration needed to provide food for the city. The second is that some types of agricultural production, notably rice production, can generate methane emissions;

management practices can be used to reduce those emissions. Other types of food production provide short-term sequestration of carbon. An important added benefit to local food production is that it supports local farmers, provides jobs, and better connects urban dwellers to the source of their food.

Transport

- Primary energy or CO_2/vehicle-km
  This indicator provides a measure of the average fleet efficiency of all vehicles in a city. Calculating this indicator requires knowing the total trip length of all public transportation modes (subway, bus, street cars) in addition to total trip length of all private transport (cars and taxis), as well as the total trip length of all trucks (light, medium, and heavy duty), as well as the total energy consumption of these vehicles. It could alternatively be estimated from a calculation of the average fleet efficiency (by vehicle type) in MJ/km along with the annual vehicle kilometers travelled by vehicle type.
- Primary energy or CO_2/person-km
  This indicator provides a measure of the energy or carbon intensity of moving people around a city. Calculating this indicator is challenging, since it requires knowing the turnover (passenger-kilometers) of all public transportation modes (buses, light rail, subway, etc), and estimating the total person-trip-kilometers for all private travel in cars and taxis, as well as the total energy consumption of these travel modes.
- Primary energy or CO_2/ton-km
  This indicator provides a measure of the energy or carbon intensity of moving goods to and around a city. Calculating this indicator is challenging since it requires knowing the total freight turnover of a city (in ton-km) and the energy consumption of the vehicles used for freight delivery.
- Kilometers of high capacity public transit systems per 100,000 population
  High capacity public transit (such as bus rapid transit, BRT) facilitates movement of large numbers of people typical of morning and evening commute hours. The more extensive the system, the higher likelihood

that commuters will choose this mode of travel over less efficient modes

- City resident public transit use (number of public transit trips per capita)
  The degree to which residents choose to ride public transportation in their total annual travel reflects both the accessibility and desirability of public transportation as well as results in lower transportation energy consumption and road congestion compared to use of personal cars. Data on total kilometers of travel in personal cars are generally not available, but surveys can provide a baseline of total personal travel by mode.
- Urban density
  Higher density land use is strongly correlated with lower per capita energy and resource consumption. For transportation, higher density allows for greater access to public transportation, reduces transit network length, and reduces the need for private cars. Urban density can be measured in terms of population per square kilometer, excluding parks and designated open space; as the number of dwelling units per square kilometer; or as a floor area ratio of the total floor area of buildings divided by the total land area used.
- Share of each city's alternative-fueled (hybrid, CNG, EV, NG) vehicles of the total vehicle fleet (government and private, buses, cars, trucks)
  City governments often own and operate a fleet of vehicles, including cars, trucks, buses, and specialized vehicles, running on gasoline and diesel. This indicator measures the number of these vehicles that run on alternative fuels with lower emissions than standard internal combustion engines.
- Two-wheeled vehicle ownership per capita
  Two-wheeled vehicles, including conventional bicycles, e-bikes, and motorcycles, offer some of the highest energy efficiencies of urban transportation modes. Registration data should be available for motorcycles and e-bikes.

Creation of a Low Carbon Development Plan

The preceding examination of indicators provides cities with a way to define and track low-carbon development. These indicators can be incorporated into a development plan or action plan. Essential steps[3] that are commonly used in the creation of a low-carbon development plan are:

1. Leadership Commitment

The first essential step is commitment by the city's leadership. With the city's attention turned to low-carbon development, and sufficient staff and time and resources committed to the effort, the city can successfully develop and implement its low-carbon plan.

2. Conduct Energy and Carbon Emissions Inventory

a) Identify the main sources of energy and carbon
b) Identify options for energy and carbon savings (rough analysis)

Two main emission sources are CO_2 and CH_4. CO_2 comes primarily from energy consumption of fossil fuels, as well as from non-energy industrial processes (like cement production), and forest loss. CH_4arises from agriculture (especially rice production), animal husbandry, other land use, industry (e.g., coal-bed methane), and waste decomposition.

The emissions inventory is a best estimate of emissions from activities in the city or province – not a precise measurement. The emissions inventory covers sources of CO_2 and CH_4 from the following sectors: electric power, industrial, residential, commercial, transportation, land management (agriculture and other land use, rural and urban), and waste.

Scope of the Inventory

Since some emission-generating activities may cross city boundaries, it is important to clearly define the scope of the emissions inventory, to know what emissions get counted by the city. Internationally-recognized inventory protocols have defined three emission scopes: (1) direct, (2) indirect, and (3) associated emissions. Table 1 explains what emissions are included under each scope.

Data Needs

City staff preparing the carbon emissions inventory must work with the local and provincial statistical bureau, with utilities supplying electricity to the city, with transportation and waste agencies, as well as enterprises. The basic emission sources and data needed are summarized in Table 2. The energy and other data on emission sources and activities, combined with emission factors, yield a GHG emissions inventory.

Table 1. GHG Emissions Inventory Scope

Emissions Scope	Scope Acitivities
Scope 1: Direct Emissions: Generated Within City Boundaries	Direct Energy Consumption within the City (fuel for Industry, Heating, Cooling, Electricity generation, Infrastructure, etc.) Transportation within the City Land Use and Waste Management within the City
Scope 2: Indirect Emissions: Due to Activities Within City Boundaries, Generated Outside City Boundaries	Import of Electricity and Heating used in the City
Scope 3: Associated Emissions: Due to City Activities, Occuring Across or Outside City Boundaries	Intra-regional Transportation City Waste in Landfills outside the City

Source: Clean Air-Cool Planet, 2010.

3. Set Targets

a) Forecast energy, carbon, and GDP under different scenarios (Business-As-Usual, Savings Scenario)
b) Set targets based on scenario forecasts
c) If time is available, set targets based on detailed analysis of potential savings, policies and measures (see next step)

Target setting involves choosing the type of target, and the target value. Targets need to be measurable and reportable, so that progress toward goals can be tracked. A physical target is preferable—such as total CO_2 emissions, or energy use, or amount of wind energy—because it can be measured and has a direct influence on the health of the city and province. Economic targets are also important. The target value is set by projecting energy and carbon in scenario analysis (Business-As-Usual Scenario, and Savings Scenario), and evaluating the impact of potential policies.

4. Create a Low-Carbon Development Plan (Climate Action Plan + Low-Carbon Economic Plan), with Policies and Actions to Meet Targets

a) Analyze and select policies and actions (detailed analysis, including co-benefits, costs, and savings)
b) Clearly state the goals for each action and how progress will be measured
c) Choose Policy Mechanisms (Action Plan) to help meet provincial targets

Table 2. Data Needs for a Greenhouse Gas Emissions Inventory

Sector	Data on emission sources
Electric Power	Energy mix and amount of generation: kWh from coal, natural gas, oil, hydro, wind, solar, nuclear, etc.
Industrial	Electricity and fuel (natural gas, coal, heat, others) consumption
Residential	Electricity and fuel (natural gas, coal, heat, others) consumption Building floor space and type
Commercial	Electricity and fuel (natural gas, coal, heat, others) consumption Building floor space and type
Transportation	Electricity and fuel (gasoline, diesel, others) consumption Mix of Transport Modes (feet, bicycle, motorbike, bus, light rail, train, auto, truck) Vehicle Efficiencies (Fuel Economy) for each mode Vehicle Miles Traveled (VMT) on local roads, for each mode VMT on highways (related to the jurisdiction)
Land Use	Hectares of food production, by type (rice, wheat, etc.) Numbers of cattle, pigs, horses Hectares of Forest cover (existing, removed, added)
Waste	Total landfill waste (tonnes) Typical composition of waste (organic matter, plastics and other non-degradable material, land-cover materials)

The savings potential from the policies will depend on that province's situation (e.g., baseline inventory, mix of efficiencies in building stock, etc.). The cost in the province will depend on that province's situation (e.g., energy pricing, renewable energy resources), as well as a typical unit cost. How can a city choose which policies it needs to meet its Target? First conduct a rough review of potential policies and actions, qualitatively considering estimates of savings and costs. Next, choose a shorter list of actions for detailed, quantitative analysis. Closely connect the actions to the Emissions Inventory and Scenarios, addressing each sector of the economy. Also consider input from research institutes, the community, businesses, and government officials.

5. Implement Policies and Actions

a) Identify and allocate responsibility
b) Set aside funding for implementation
c) Set timetables
d) Support policies with incentives, penalties, training and public outreach

6. Monitor, Report and Verify Progress

Progress must be tracked with Monitoring, including Reporting and Verification. Reporting on intensity must include data on energy use, carbon emissions, and data on economic activity, to verify the resulting intensity number. Public reporting of data, along with progress toward goals, focuses attention and effort from government, enterprises, and the public, and helps to achieve the targets. City government websites are an effective means for publicly tracking progress on energy, carbon, and low-carbon economic development.

The guidance here focuses on the city level; similar steps can be undertaken at every level, from enterprise, to city, province, and country.

Policies and Actions to Achieve Low Carbon Development

Because the heart of a low-carbon development plan is its actions, the guidebook provides an extensive list of policy options and performance indicators. International experience demonstrates that end-use energy efficiency has the capability to significantly reduce GHG emissions at low cost. Because energy and economic structures vary from city to city, the impact of the policies and associated costs need to be evaluated in order to determine priorities and to assist the local governments to select most cost effective policies. This report intends to provide basic indicators and methods of policy choice by categorizing them into "High, "Medium", "Low" in terms of the energy and carbon savings, and similar categorization in terms of implementation costs. Examples of such policy categories and their significance in the building and industry sectors are shown in Table 3 and Table 4, and Figure 1 and Figure 2, respectively. The tables list the policy options, noting the rough scale of energy and carbon savings, as well as costs to implement. The figures highlight the significance of policy options by graphically ranking them. A quantified cost-benefit analysis of policies would further facilitate policy prioritization and implementation by local governments. More detailed analysis also requires more public datasets and survey results. The guidebook includes policies and actions to achieve low-carbon growth in the following sectors: cross-cutting (not focused on a specific end-use sector), industry, buildings, transportation, power, agriculture and forestry. For each policy or action identified in each sector, the following information is included, where available:

- Description
- Performance metric
- GHG emission reduction potential
- Cost-effectiveness

The policies draw on international examples, including both national and state or provincial level measures. Chinese approaches are also included if they are considered to be successful or innovative. An excerpt from one policy for the industrial sector – target setting – is provided below to illustrate the format and content of the policies included in the guidebook.

TARGET SETTING[4]

Policy Description

Target-setting for energy efficiency or GHG emissions reduction is a common practice; a recent survey identified 23 such programs in 18 countries around the world, including countries in Europe, the U.S., Canada, Australia, New Zealand, Japan, South Korea, and Chinese Taipei (Taiwan) (Price, 2005). Targets are typically either voluntary commitments or negotiated agreements, but can also be mandatory targets assigned by the government as in the case of China's Top-1000 Energy-Consuming Enterprises Program. Targets can be agreed upon by individual companies or by industrial sectors through organizations such as industrial associations. Voluntary commitments are often made by companies either individually or through government programs. The U.S. Environmental Protection Agency's (EPA's) Climate Leaders is comprised of approximately 200 companies that have committed to undertake a corporate-wide inventory of their GHG emissions, set aggressive reduction targets, and report their progress annually to the EPA (U.S. EPA, 2010a). Voluntary commitments are also made by industrial sectors such as the

Cement Sustainability Initiative (CSI) of the World Business Council for Sustainable Development (WBCSD) (CSI, n.d.; WBCSD, 2010) and the International Aluminium Institute (IAI, 2009).

Performance Metrics

There are three performance metrics for target-setting programs:

1) The number of enterprises with target-setting contracts
2) The number of enterprises that meet or surpassed their targets
3) The average savings per participating enterprise

GHG Emission Reduction Potential

GHG emissions reductions by U.S. EPA Climate Leaders companies are estimated to be equivalent to more than 50 $MtCO_2e$ per year (U.S. EPA, 2010b). Given that there are 195 companies currently in the program, this is an annual average emissions reduction of approximately 256,000 tCO_2e per company.

The UK Climate Change Agreements have also resulted in large CO_2 emissions reductions. During the first target period (2001-2002) total realized reductions were nearly three times higher than the target for that period (Future Energy Solutions, 2004). Industry realized total reductions that were more than double the target set by the government during the second target period and that were nearly double the target during the third and fourth target periods (AEA Energy & Environment, 2009; DEFRA, 2005; DEFRA, 2007; Future Energy Solutions, 2005). There are approximately 9000 facilities participating in this program. If the annual emissions reductions of 20.3 $MtCO_2$ achieved in 2007 and 2008 are divided evenly among these 9000 plants, the average emissions reductions are 2255 tCO_2/plant/year.

In November 2009, China's National Development and Reform Commission announced that the Top-1000 program had surpassed its target energy savings of 100 Mtce, saving 106 Mtce by the end of 2009 (NDRC, 2009). Dividing this savings over 5 years by the roughly 1000 participating enterprises results in annual savings of 20,000 tce per plant. Using a conversion factor of 2.5 tCO_2/tce results in estimated average annual per plant savings of 5,000 tCO_2.

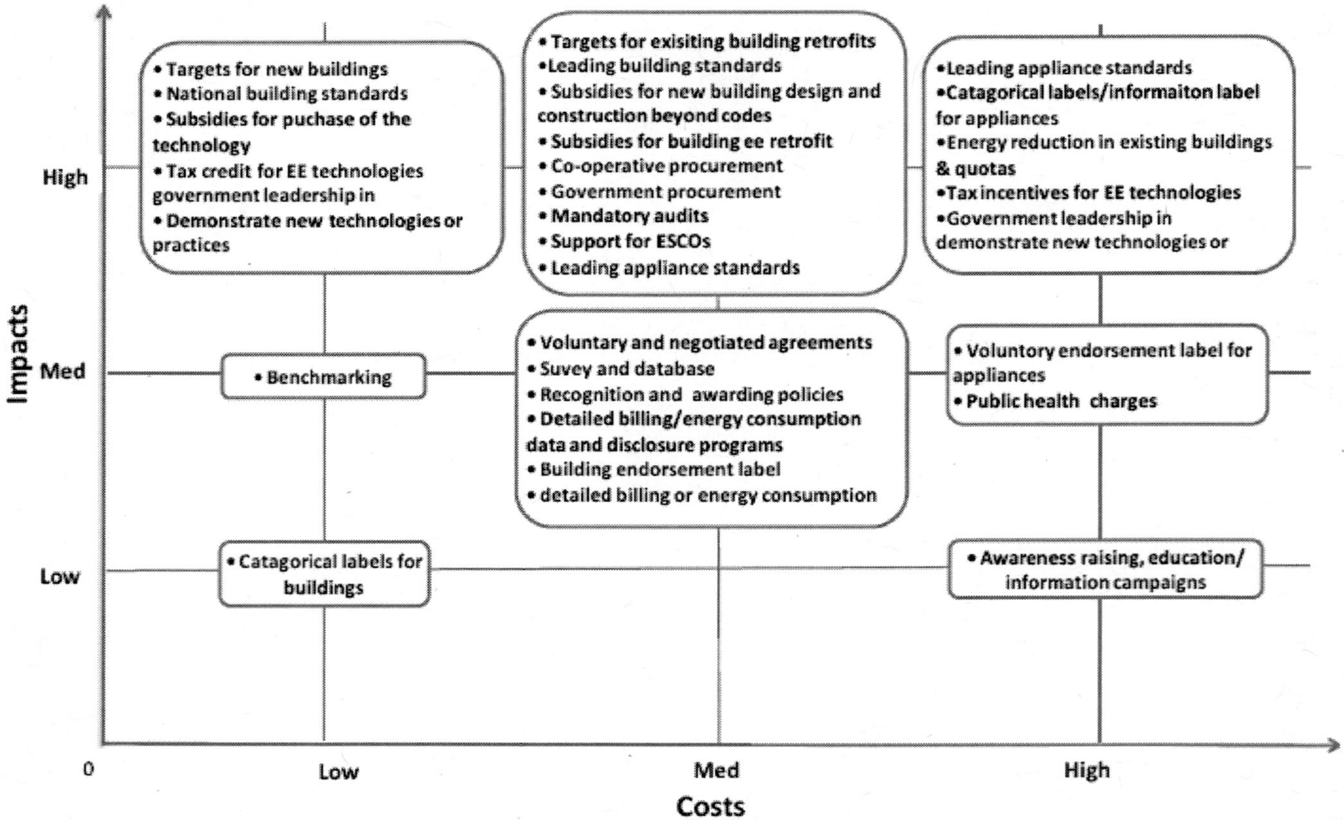

Figure 1. Costs and Savings of Energy Efficiency Policies for Buildings and Appliances.

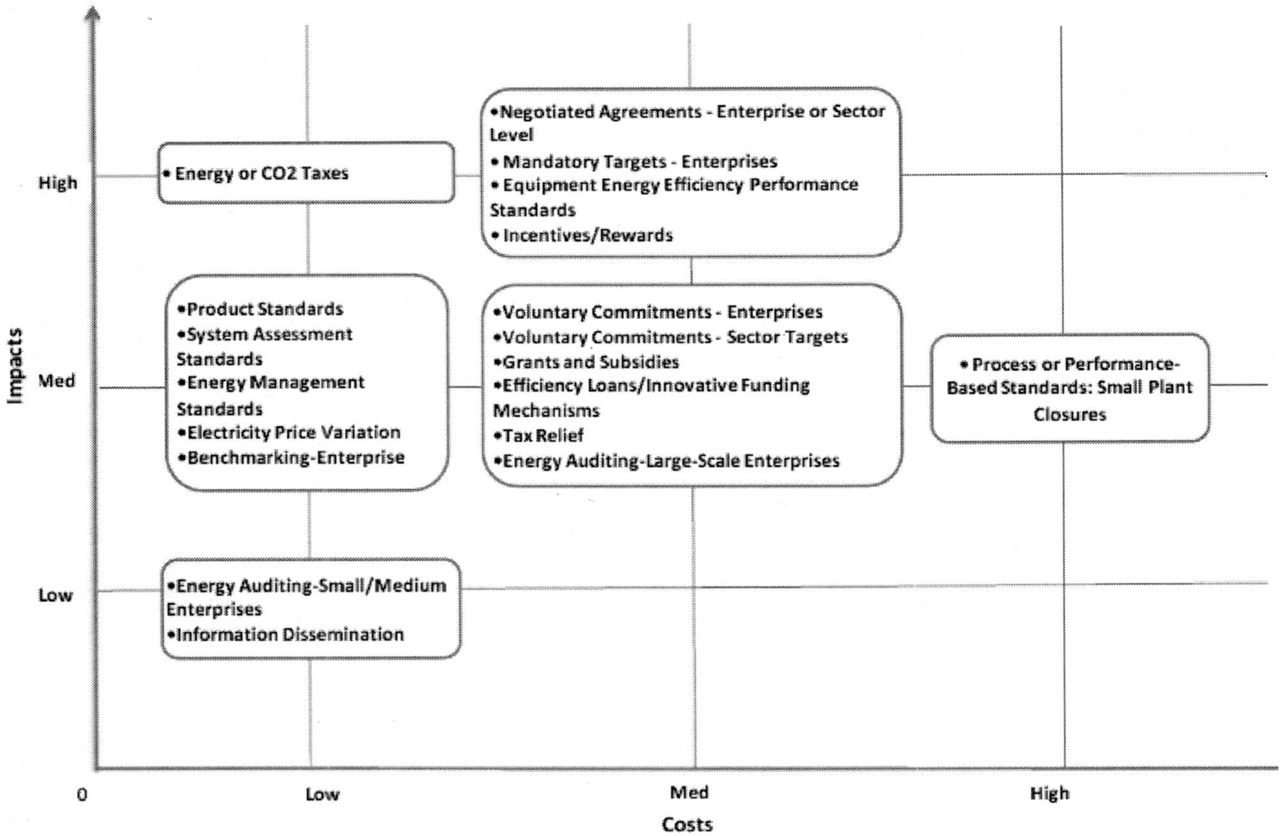

Figure 2. Costs and Savings of Energy Efficiency Policies for Industry.

Table 3. Examples of Policies and Programs for Building Sector

Policy/Program Identification		Cost and Impact						
Policy Option	Performance Metric	GHGs Reduction by 2020			COST			Notes
		Hi	Med	Lo	Hi	Me d	Lo	
Targets								
Targets for new buildings	Inspection and evaluation on compliance	X					X	both at design and construction phase
Targets for existing building retrofit	m2 retrofitted	X				X		
Standards								
National Building Standards	Level of building codes; compliance level	X					X	
National Appliance Standards	Level of standard; compliance level	X				X		
Certification, Labels, Voluntary Programs								
Appliance Categorical/information label	Level of compliance; product grade market shift	X			X			
Voluntary endorsement label for appliances and buildings	Level of compliance; product grade market shift		X		X			effective with financial incentives, voluntary agreements and regulations
Energy management								
Energy reduction in existing buildings	Coverage; compliance level	X			X			

Policy/Program Identification		Cost and Impact						
Policy Option	Performance Metric	GHGs Reduction by 2020			COST			Notes
		Hi	Med	Lo	Hi	Me d	Lo	
EE Technology/measure Promotion								
Subsidies for building ee retrofit	Retrofitted area	X				X		
Tax credit/incentives for EE technologies	Increased sales in EE technologies	X			X		X	low cost for tax credit
Setting technology dissemination goals	MW installed							
Zoning								
Mixed-use Zoning; green space zoning	Area regulated; stringency of requirements	X						
Public Sector Leadership								
Government procurement	clear and accessible information, compliance level	X				X		
Building Commissioning/Auditing								
Mandatory audits	Number of audits conducted	X				X		most effective if combined with other measures such as financial incentives

Table 3. (Continued)

Policy/Program Identification		Cost and Impact						
Policy Option	Performance Metric	GHGs Reduction by 2020			COST			Notes
		Hi	Med	Lo	Hi	Me d	Lo	
Reporting								
Detailed billing or energy consumption data and disclosure programs	Data availability		X				n. a	success conditions: combination with other measures and periodic evaluation.

Table 4. Policies and Programs for Industry Sector

Policy/Program Identification		Cost and Impact					
Policy Option	Performance Metric	GHGs Reduction by 2020			Cost		
		Hi	Med	Lo	Hi	Med	Lo
Targets							
Voluntary Commitments - Enterprises	Average savings per participating enterprise; # of enterprises with targets; # of enterprises that meet or surpassed targets		X			X	

Policy/Program Identification		Cost and Impact					
Policy Option	Performance Metric	GHGs Reduction by 2020			Cost		
		Hi	Med	Lo	Hi	Med	Lo
Voluntary Commitments - Energy-Saving and GHG Emission Reduction Sector Targets	Achieved savings/emissions reductions		X			X	
Negotiated Agreements - Enterprise or Sector Level	Average savings per participating enterprise; # of enterprises with targets; # of enterprises that meet or surpassed targets	X				X	
Mandatory Targets - Enterprises	Average savings per participating enterprise; # of enterprises with targets; # of enterprises that meet or surpassed targets	X				X	
Standards							
Product Standards	Energy saved and/or CO_2 emissions reduced annually		X				X
System Assessment Standards	Energy saved and/or CO_2 emissions reduced annually		X				X
Process or Performance-Based Standards: Equipment Energy Efficiency Performance Standards	Sector reaches "advanced minimum"	X				X	

Table 4. (Continued)

Policy/Program Identification		Cost and Impact					
Policy Option	Performance Metric	GHGs Reduction by 2020			Cost		
		Hi	Med	Lo	Hi	Med	Lo
Process or Performance-Based Standards: Small Plant Closures	Final/primary energy saved per t cement; final/primary energy saved per t iron; final/primary energy saved per t steel; electricity saved per kWh; final/primary energy saved per t paper; final/primary energy saved per t aluminum		X		X		
Energy Management Standards	Information on standards disseminated to industry; standards adopted		X				X
Fiscal/Financial Instruments							
Energy or CO_2 Taxes	Benefit net of costs per ton CO_2 saved	X					X
Grants and Subsidies	Energy saved and/or CO_2 emissions reduced per unit of funding provided		X			X	
Energy Efficiency Loans and Innovative Funding Mechanisms	Energy saved and/or CO_2 emissions reduced per unit of funding provided		X			X	
Tax Relief	Energy saved and/or CO_2 emissions reduced		X			X	
Electricity Price Variation			X				X

Policy/Program Identification		Cost and Impact					
Policy Option	Performance Metric	GHGs Reduction by 2020			Cost		
		Hi	Med	Lo	Hi	Med	Lo
Incentives/Rewards	Energy saved and/or CO_2 emissions reduced	X				X	
Energy Auditing							
Large-Scale Enterprises	# energy audits conducted; typical savings identified/audit		X			X	
Small and Medium Enterprises	# energy audits conducted; typical savings identified/audit			X			X
Benchmarking							
Enterprise Level	# enterprises undertaking benchmarking; Energy saved and/or CO_2 emissions reduced as a result of benchmarking		X				X

Cost-Effectiveness

A 2004 evaluation of the first Dutch Long Term Agreements (LTA1s) calculated that the cost of the LTA1s was about \$10 per tonne of CO_2 reduced, assuming the savings last for 10 years and using a social discount rate of 5% (Blok et al., 2004). Sweden's program for improving energy efficiency in energy-intensive industry was introduced after the adoption of an electricity tax in 2004. In 2006, 98 companies submitted reports outlining nearly 900 energy-efficiency improvements that they plan to undertake by 2009 which will cost the companies about €110 million and reduce electricity consumption by 1 TWh/year, saving companies €55 million per year. In addition, the companies will receive €17 million tax reductions through this program (SEA, 2005; SEA, 2006; SEA, 2007).

In 2007, the UK's National Audit Office found that in general the benefits of the Climate Change Agreements (CCAs) outweighed the program administrative costs (NAO, 2007). An independent evaluation of the CCA program found that in addition to the energy and GHG emissions reductions, the program provided "positive macroeconomic effects in economic terms, with small increases in GDP and employment" (Barker et al., 2007). It is estimated that UK industry saves over \$800 million per year as a result of meeting the UK CCA targets (Pender, 2005), an average annual savings of over \$90,000 per year for the approximately 9000 participants. Another analysis estimated that the benefit net costs per ton of carbon saved in 2010 from the CCAs will bc \$38/tCO_2 (DEFRA, 2006).

CONCLUSION

Much work lies ahead to appropriately define and implement low-carbon development at the city level. China has announced the goal to achieve lower carbon intensity and to develop low carbon demonstration cities. Even so, there is still a strong need for methodologies, measures, and tools to achieve these goals. This paper, and the guidebook upon which it is based, provides an important foundation for these efforts. The low carbon indicators examined here provide clear metrics for tracking energy and carbon savings over time, as well as comparing progress among cities. The planning steps outlined in the paper can help cities shape a comprehensive effort and aim for climate-friendly city development. Finally, the policy options and categorization illustrated in this paper provide guidance for cities to take action.

REFERENCES

Clean-Air Cool-Planet. 2010. Community Toolkit. http://www.cleanaircoolplanet.org/for_communities/toolkit_sitemap.php. Accessed November, 2010.

International Council for Local Environmental Initiatives (ICLEI) – Local Governments for Sustainability.

2009. U.S. Mayors Climate Protection Agreement Climate Action Handbook. Available at: http://iclei-usa.org/mount_iclei/iclei/action-center/planning/climate-action-handbook

Newsham, G., Mancini, S., and Birt, B., 2009. Do LEED-certified buildings save energy? Yes, but ~ National Research Council Canada – Institute for Research in Construction. Available at: http://www.nrc-cnrc.gc.ca/obj/irc/doc/pubs/nrcc51142.pdf

Price, L., Jiang, Y., Worrell, E., Du, W., Sinton, J.E., 2003. Development of an Energy Conservation Voluntary Agreement Pilot Project in the Steel Sector in Shandong Province: Report to the State Economic and Trade Commission, People's Republic of China. Berkeley, CA: Lawrence Berkeley National Laboratory (LBNL-51608). English version available at: http://china.lbl.gov/sites/china.lbl.gov/files/LBNL 51608.2003. English.pdf

Chinese version available at: http://china.lbl.gov/sites/china.lbl.gov/files/LBNL51608.2003.Chinese.pdf

Renewable Energy and Energy Efficiency Partnership (REEEP), Alliance to Save Energy (ASE) and American Council on Renewable Energy (ACORE). 2010. Compendium of Best Practices: Sharing Local and State Successes in Energy Efficiency and Renewable Energy from the United States. Available at: http://www.reeep.org/file_upload/5272_tmpphp4EtwHc.pdf

Roland-Holst, D., 2008. Energy Efficiency, Innovation, and Job Creation in California. Berkeley, CA: University of California, Berkeley. Available at: http://www.next10.org/pdf/report_eijc/75_01-2%20ClimateAction_Report%2004B_online.pdf

End Notes

[1] The proposed activities are guidelines for cities to develop the plans and it is unclear what measures will be adopted. In general, policies such as incentives, standards and labeling scheme, quotas have all been proven effective both internationally and in China.

2 Note that a comprehensive emissions inventory would include all *six* GHGs recognized under the Kyoto Protocol: CO_2, CH4, N_2O (from fertilizers and manure), SF6 (from electrical systems, magnesium production), HFCs (refrigeration, semiconductor manufacturing, aluminum smelting), and PFCs (aluminum and semiconductor production).

[3] Based on steps developed by ICLEI; for example, ICLEI 2009.

[4] See the guidebook for the references provided in this section.

In: Going Green in China: Policy and Potential ISBN: 978-1- 62100-691-6
Editors: Spencer B. Talscott

Chapter 3

THE GREENING OF THE MIDDLE KINGDOM: THE STORY OF ENERGY EFFICIENCY IN CHINA*

Mark D. Levine, Nan Zhou and Lynn Price

DISCLAIMER

This document was prepared as an account of work sponsored by the United States government. While this document is believed to contain correct information, neither the United States government nor any agency thereof, nor The Regents of the University of California, nor any of their employees, makes any warranty, express or implied, or assumes any legal responsibility for the accuracy, completeness, or usefulness of any information, apparatus, product, or process disclosed, or represents that its use would not infringe privately owned rights. Reference herein to any specific commercial product, process, or service by its trade name, trademark, manufacturer, or otherwise, does not necessarily constitute or imply its endorsement, recommendation, or favoring by the United States government or any agency thereof, or The Regents of the University of California. The views and opinions of authors expressed herein do not necessarily state or reflect those of the United States

* This is an edited, reformatted and augmented version of the Ernest Orlando Lawrence Berkeley National Laboratory publication, LBNL-2413E, dated May 2009.

government or any agency thereof, or The Regents of the University of California.

Ernest Orlando Lawrence Berkeley National Laboratory is an equal opportunity employer.

INTRODUCTION

The dominant image of China's energy system is of billowing smokestacks from the combustion of coal. More heavily dependent on coal than any other major country, China uses it for about 70 percent of its energy (NBS, 2008). Furthermore, until recently, China had very few environmental controls on emissions from coal combustion; recent efforts to control sulfur dioxide (SO_2) emissions appear to be meeting with some success (Economy, 2007, 2009).

Figure 1 shows the dominant use of coal in China's energy system from 1950 to 1980 (NBS, various years). However, this is just one side of China's energy story. Figure 2 illustrates the second part, and what may be the most important part of the story—China's energy system since 1980, shortly after Deng Xiaoping assumed full leadership. This figure compares the trends in energy consumption and gross domestic product (GDP) by indexing both values to 100 in 1980. The upper line shows what energy consumption in China would have been if it had grown at the same rate as GDP, since energy consumption usually increases in lockstep with GDP in an industrializing, developing country, at least until it reaches a high economic level.

The lower line in Figure 2 shows China's actual energy consumption, also indexed to 1980. The striking difference between the lines shows that GDP in China grew much faster than energy demand from 1980 to 2002. As a result, by 2002 energy and energy-related carbon dioxide (CO_2) emissions were more than 40% percent of what they would have been if energy and GDP had grown in tandem.

In the next chapter of China's energy history, from 2002 to 2005, the increase in energy demand outstripped a very rapidly growing economy, and because of the large size of the Chinese economy, the increase had substantial impacts. The construction of power plants increased to 100 gigawatts per year; over the three-year period newly constructed plants had a capacity of more than 30 percent of total electricity-generation capacity in the United States. At the same time, energy-related CO_2 emissions in China increased dramatically.

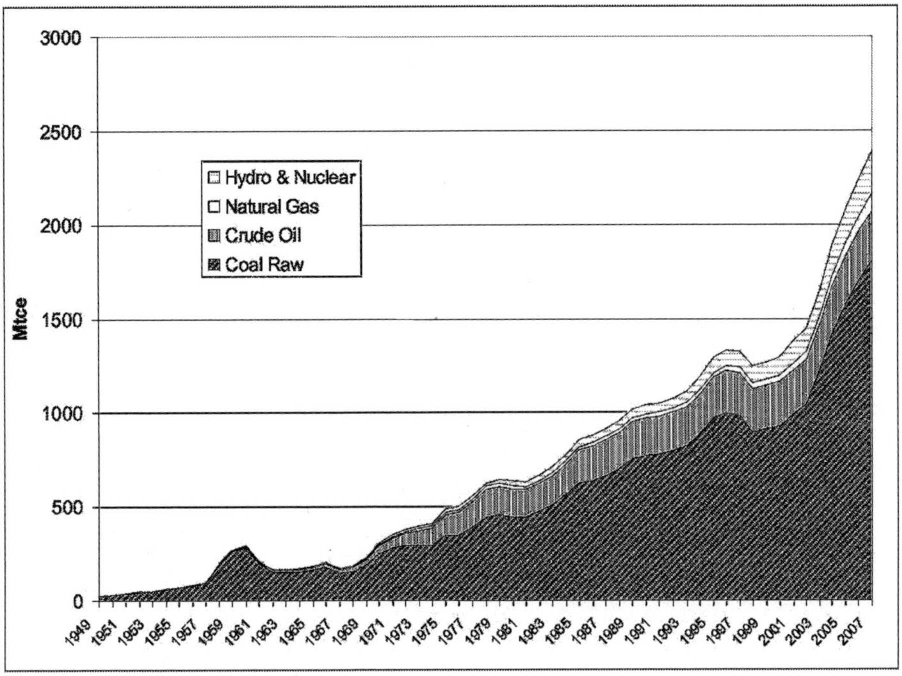

Figure 1. Coal dominates energy consumption in China. Source: NBS, various years.

In the latest stage, another abrupt change, this time for the better in terms of energy efficiency, began late in 2005. As senior officials in the government turned their attention to the problem of growing energy demand, the government set a mandatory goal for 2010 of a 20 percent reduction in energy intensity (defined as energy use per unit of GDP) from 2005 levels. To meet this goal, China undertook significant legislative, regulatory, and organizational reforms at the national, provincial, and municipal levels to ensure that measures to reduce energy intensity would be implemented in all sectors and activities in China.

At the time of this writing, it appears that China is on its way to meeting the 20 percent goal, thus reducing CO_2 emissions by 1.5 billion tones, as compared with consumption at 2005 energy-intensity levels. In this paper, we describe and assess these three significant periods in China's energy story and provide a context by briefly reviewing the three decades prior to 1980.

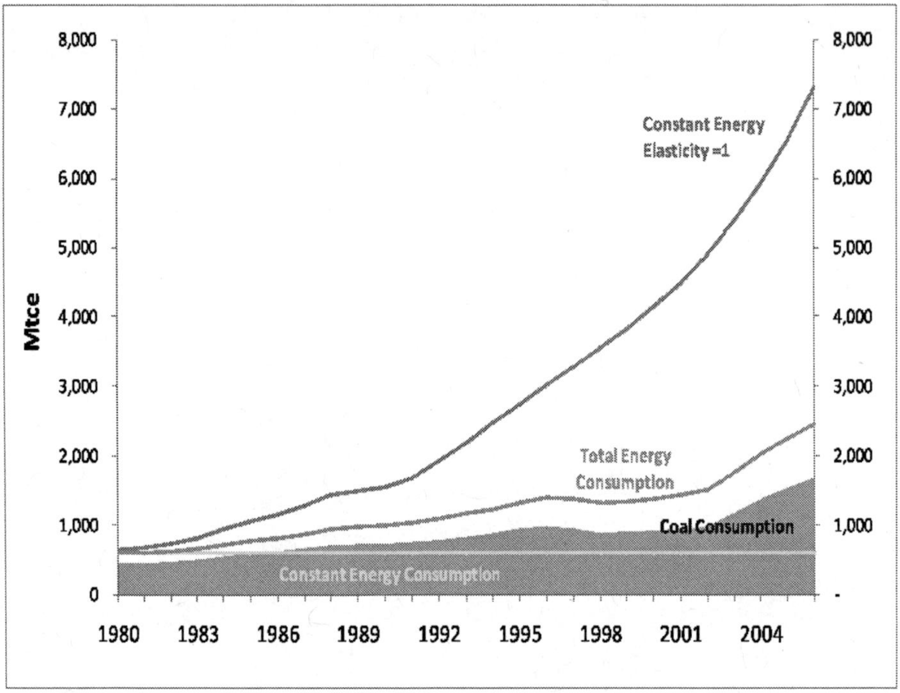

Figure 2. Actual energy demand in China is very much lower than energy demand at constant energy intensity, 1980-2006. Source: NBS, various years.

The Soviet Model: 1949 through 1980

From the beginning of the communist regime in 1949 until the ascendancy of Deng Xiaoping in 1979, China's energy policy and the system it created followed the Russian model— rapid increases in energy supply, low energy prices, centralized energy allocation to provide energy to heavy industry, and a disregard for environmental effects. The result of this policy was one of the fastest growing and least efficient energy systems in the world, on both the supply side and demand side (Figure 3a).

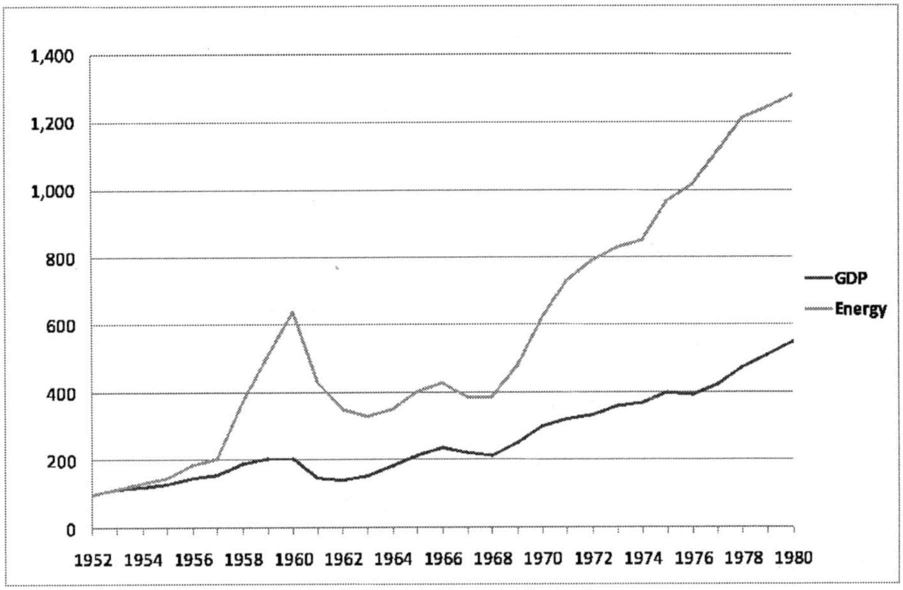

Figure 3a. Energy demand grew twice as fast as GDP from 1952 to 1980. Source: NBS, various years.

THE "CLASSIC" PERIOD OF ENERGY EFFICIENCY: 1980 THROUGH 2002

In 1980, in response to serious concerns in the academic community about Chinese energy policy, Deng Xiaoping adopted a strategy of reducing energy intensity. His stated goal was to quadruple GDP while only doubling energy consumption over the 20-year period, 1980–2000 (Lu, 1993).

China exceeded this goal both in the increase in GDP and the reduction of energy intensity (Figure 3b). This was achieved through a variety of innovative policies and programs, many of which were developed by the Chinese without any significant knowledge of what other countries had done to promote energy efficiency. Not until a decade after China had embarked on its program to reduce the energy intensity of its entire economy did officials begin to establish ties with the energy-efficiency community outside its borders. Two of these policies—one on investment in energy efficiency and the other on establishing centers of expertise in energy efficiency throughout the nation—were far ahead of their time. To this day, no other country has effectively implemented such policies at the level done in China.

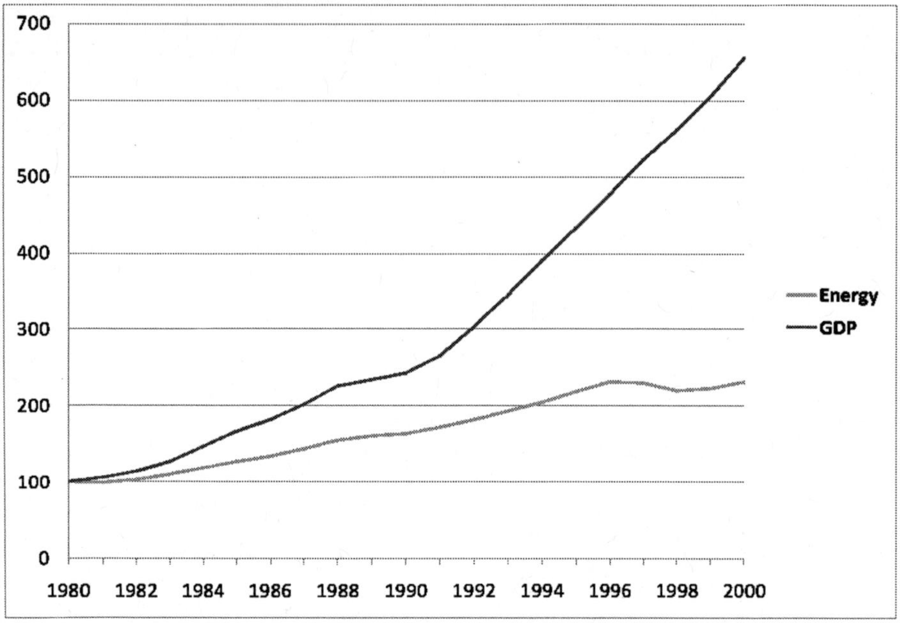

Figure 3b. Energy demand grew 50 percent faster than GDP from 1952 to 1980, albeit from a very low level. Source: NBS, various years.

As shown in Figure 4, investment in energy efficiency accounted for more than 10 percent of total energy investment in 1981, the first year of Deng Xiaoping's program.[1] Investment later increased to 12 percent before slowly declining to a sustainable level of 5 to 6 percent.

In the early and mid-1980s, energy efficiency could be achieved inexpensively by fixing leaky pipes, inefficient boilers, and other equipment and by changing sloppy energy-management practices. Thus a 10 percent investment led to a much larger reduction in the increase in energy demand than a comparable investment led to increased supplies. The investment program alone—which was just one of a number of energy-efficiency policies— achieved a significant portion of Deng Xiaoping's goals.[2]

The investment program spurred the development of new institutions, such as the China Energy Conservation Investment Corporation, which developed branches throughout the country to channel investments into energy efficiency and co-generation (strongly supported by the Chinese government). At the national level, the Chinese created the Bureau of Energy-Saving and Comprehensive Energy Utilization in the State Planning Commission (SPC). Today, after various restructurings, SPC has become the National

Development and Reform Commission (NDRC). NDRC and its forerunner commissions are half a level above ministries in the Chinese hierarchy. All major requests to the State Council from ministries are supposed to—and often do—flow through NDRC. The very existence in the 1980s of a bureau for energy efficiency at this level indicates its importance.

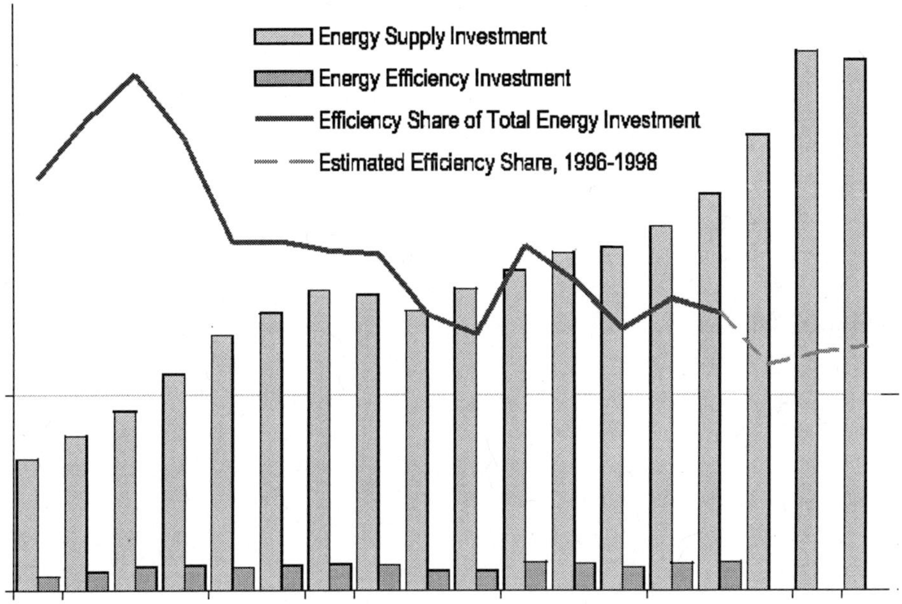

Figure 4. Investment in energy efficiency became a significant part of total energy investment in 1981. Source: NBS, various years.

This bureau created a variety of programs to promote energy conservation (a term for energy efficiency that is still often used in China). One of the most significant of these new policies was the establishment of energy-conservation service centers throughout the country. At their peak, there were more than 200 of these centers, employing more than 7,000 people across China. For a more complete description of institutional reforms to promote energy efficiency see Sinton and Levine (1998).

It is instructive to ask what might have happened if China had not embarked on such an aggressive and innovative policy. As Figure 2 shows, Chinese actions going back to 1980 enabled the country to avoid a situation in which global energy-related CO_2 emissions in 2007 would be three times higher than they are. This would have resulted in global emissions in 2007 at levels projected by the International Energy Agency for 2025 (IEA, 2008).[3]

OUT-OF-CONTROL GROWTH IN ENERGY DEMAND: 2002 THROUGH 2005

In spite of significantly slower increase in energy demand compared to GDP, there were signs in the late 1990s that energy efficiency was becoming less important to Chinese policy makers. Funding for the government effort to gather and analyze energy data was reduced significantly throughout the decade; as a result, the data were not only less comprehensive, but also less accurate. China's system for gathering data on energy consumption, which had been one of the best among developing nations, was much weakened by the end of the 1990s. However, data on energy supply, which come from a small number of energy-supply companies and are relatively easy to track, remained plentiful.

By the turn of the century, little attention was paid to energy efficiency at the industrial-enterprise level. By law, all key industries (i.e., industries that consume more than 10,000 tons of coal equivalent per year) were required to have an energy manager. However, by 2000, many large enterprises had energy managers in name only, if at all. This meant that the enterprises consuming the most energy had lost the expertise (and often the data) to assess and improve their energy efficiency.

Other signs that energy efficiency was being given lower priority included the decline of many of the energy conservation centers; a dispirited bureaucracy in the bureau and division responsible for energy efficiency at the central government level; reduction in budgets for energy efficiency; and most important, the lack of authority and/or willingness in national, provincial, municipal, and local government bodies to enforce laws and regulations intended to save energy.[4]

At the same time the Chinese government lost its focus on energy efficiency, China's accession to the World Trade Organization (WTO) had the inadvertent effect of undermining more than two decades of efforts to reduce energy intensity. Beginning in the early 2000s, and coinciding with China's membership in the WTO, there was a very rapid increase in exports, supported by rapid growth in industry to feed the export markets, which was heavily weighted toward energy-intensive products (i.e., products whose manufacture requires large amounts of energy and results in substantial CO_2 emissions) (Andrews-Speed, 2009).

In addition, China's internal demand for energy-intensive commodities like cement and steel to build infrastructure and cities to serve its rapidly urbanizing population outpaced the growth of less energy-intensive industries, also leading to increased overall energy use and energy-related CO_2 emissions.

These and other factors that contributed to the enormous output of energy-intensive industries in China resulted in energy demand that increased at breakneck speed and far exceeded the trajectory that would achieve the goal of quadrupling GDP while doubling energy demand from 2000 to 2020 (Figure 5).

Thus, as China developed its infrastructure, urbanized, and became the supplier of countless products to the world, its CO_2 emissions increased rapidly and dramatically. Clearly, the increase in CO_2 emissions was partially due to demand in importing nations that used the products and manufactured in China. But the commonly accepted system of accounting for greenhouse gas emissions (i.e., the convention used to assess compliance with the Kyoto treaty of Annex I countries) attributes emissions to the country that produces products rather than to the country that imports them. Indeed, changing the attribution of greenhouse gas emissions from production to consumption would be difficult because of considerable uncertainties in assessing embodied energy in products.

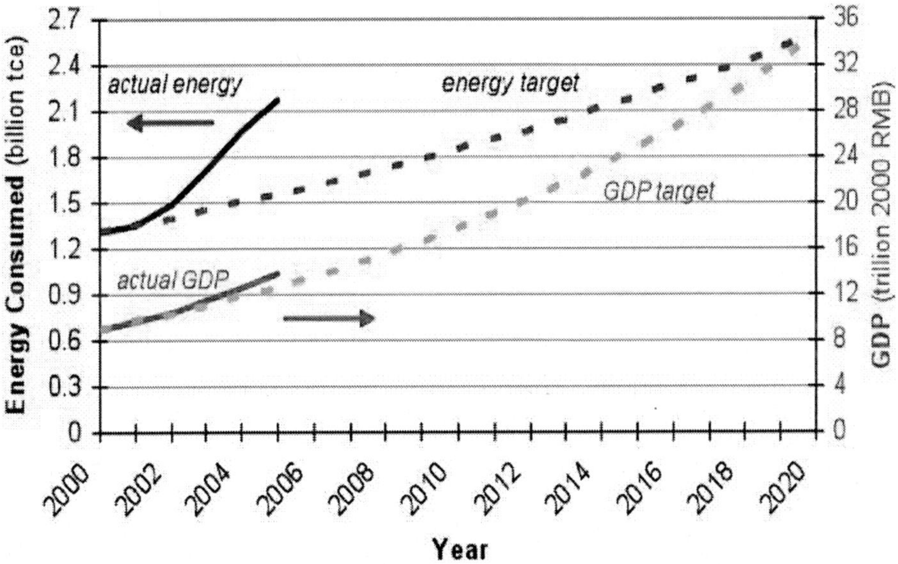

Figure 5. Beginning in 2002, increasing energy demand was on a trajectory to dramatically exceed the energy-reduction goals for 2020. Source: Lin, 2007.

In fact, in this three-year period, China's energy-related CO_2 emissions—which were 60 percent of those of the United States in 2002—approached the U.S. level by 2005 and surpassed them a year later (Figure 6) (Levine and Aden, 2008). Previous forecasts as late as 2004 by Chinese government research institutes, international agencies, and mainstream analysts from various countries had indicated that China was unlikely to overtake the United States in energy-related CO_2 emissions until 2015, or even 2020 (Levine and Aden, 2008). However, developments during the 2002 to 2005 period resulted in this wholly unexpected (and undesired) result for China.

How did this dramatic change come about, and what can the Chinese do about it? In brief, there were four major factors behind the dramatic rise in energy consumption in China:

- a gradual decline in the gathering of information on energy demand and in regulating demand by government institutions
- a rapid increase in the production of energy-intensive products for export as a result of China's membership in WTO
- increasing wealth and prosperity of a large portion of China (especially in the eastern provinces) and the associated construction of buildings and infrastructure, including transportation corridors (e.g., highways and canals for shipping freight and moving water) to serve this population
- ongoing rapid migration of people from rural areas, where they had consumed little energy, to urban areas, where energy consumption is much higher

One of the most dramatic indicators of this energy-intensive period is the increase in the production of cement. Figure 7 shows that China produces 50 percent of the cement produced worldwide (USGS, 2009). Although this is a startling statistic, it logically reflects the extraordinary pace of construction in China, which is a substantial portion of total world construction.

A Modern Re-Enactment of the Early Days: 2005 to the Present

By 2005, officials at senior levels in the Chinese government had recognized that the rapid increase in energy demand presented serious problems and was unsustainable. The rate of construction of energy-supply infrastructure—hydroelectric facilities and power plants—was putting great pressure on China's industrial system and creating difficult problems with the safety and reliability of these complex systems.

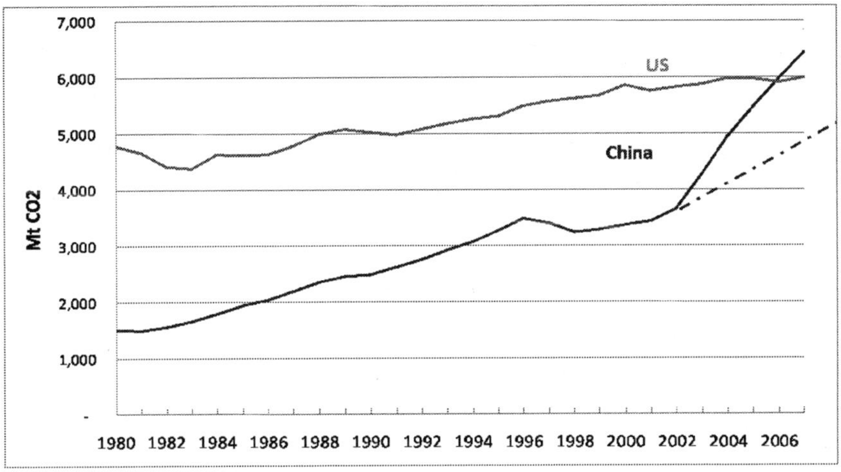

Figure 6. With the enormous increase in energy demand from 2002 to 2005, China became a larger emitter of energy-related CO2 than the United States. Source: Levine and Aden, 2008.

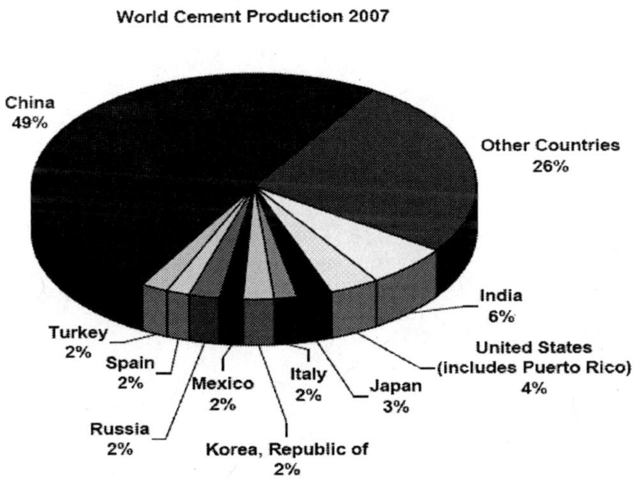

Figure 7. Cement production in China is 50 percent of world production. Source: USGS, 2009.

As measured in investment cost per unit of industrial output, energy supply is one of the most capital-intensive industrial activities. Demand for capital to build new energy-supply and conversion systems is now in competition with the demand for capital to promote balanced economic and social development. In addition, senior leaders of the Chinese government and party were becoming increasingly concerned about the negative environmental effects of the rapid increase in energy supply to meet the burgeoning demand.

In November 2005, the Politburo issued a highly unusual statement setting a mandatory 20 percent reduction by 2010 in energy intensity (measured as energy consumption per RMBY[5] of GDP). The Politburo typically addresses broad issues and sets quantitative goals for the Chinese economy as a whole; setting goals for specific industry sectors is usually the purview of the government, rather than the party. Clearly, the announcement of a 20-percent decrease in energy-intensity production indicated that the senior leaders perceived the energy problem to be extremely serious.

The level of activity that followed resembled the energy-efficiency initiatives of the early days that had led to the creative energy-conservation goals in the 6th Five-Year Plan and the intense activities that followed: legislation, regulations, government reorganization, and the creation of new institutions at the national, provincial, and municipal levels.

Nevertheless, there were important differences between the situation in 1980 and the situation in 2005. The economy in 2005 was approximately 10 times larger than in 1980 and was, therefore, much more difficult to manage. In 1980, most major energy-supply enterprises were government owned, while in 2005, although the government still exerted influence and retained some control over them, it did not own them.[6] The key difference, however, was on the demand side. By 2005, the government's ability to shape or restrain or, in many cases, even influence the demand for products, buildings, and services and the concomitant demand for energy services, was greatly limited.

In spite of these differences, the intensity and creativity of policy development for energy efficiency resembled the activity in China in the early 1980s, as well as the activity that followed the oil embargo in the United States and other industrialized countries. By 1975, three years after the embargo, many affected countries had enacted and were enforcing a variety of laws and regulations requiring or promoting energy efficiency in automobiles, buildings, and industry. The United States, Japan, and major economies in Europe had created new institutions in both the public and private sectors to carry out these laws and regulations and, for the first time, had provided government funding to promote energy efficiency.

Table 1 summarizes some of the key laws, regulations, and programs put into place in China since the Politburo directive of November 2005. These policies include the Top 1,000 Energy-Consuming Enterprises Program (Top 1,000 Program); the Ten Key Projects; allocations of government funds to support private investment in energy efficiency and pollution abatement; the creation of new government organizations and the strengthening of existing ones responsible for the design and implementation of energy-efficiency measures; and a variety of laws, regulations, and tax incentives.[7]

Many of these efforts are associated with the 11th Five-Year Plan, but some go well beyond it. Three of the most important policies are briefly described below.[8]

Ten Key Projects

In preparation for the intense focus on energy-efficiency policy that began with the November 2005 announcement by the Politburo, NDRC initiated the "Ten Key Projects," in 2004, and in 2005, the "Ten Key Projects" were incorporated into the 11th Five Year Plan. The four most significant of these projects are: the renovation of coal-fired industrial boilers; district-level combined heat and power projects; oil conservation and substitution; and energy efficiency and conservation in buildings. The expected impact of these four projects is up to 250 million tonnes carbon equivalent (Mtce) per year or about 40 percent of the 2010 target for energy intensity in China (NDRC, 2004)[9]. Twenty provincial energy-conservation centers (ECCs) were given financial support from the central government to assist in the implementation of these projects.[10]

Top 1,000 Energy-Consuming Enterprises Program

Launched in April 2006, the Top-1,000 Program was designed to improve industrial energy efficiency by targeting China's 1,000 highest energy-consuming enterprises, which account for almost 50 percent of total industrial-sector energy consumption and 30 percent of total energy consumption in China. The Top 1,000 enterprises are in nine sectors: iron and steel, petroleum and petrochemicals, chemicals, electric power, nonferrous metals, coal mining, construction materials, textiles, and paper.

Table 1. Recent key energy policies supporting China's 20% intensity reduction goal Sources: (ref)

Energy Policies	Date Effective	Responsible Agency
Fuel Consumption Limits For Passenger Cars	2004	
Medium and Long-Term Plan for Energy Conservation	2005	National Development and Reform Commission (NDRC)
Renewable Energy Law	2005	
Government Procurement Program	2005	NDRC and Ministry of Finance (MOF)
National Energy Efficient Design Standard for Public Buildings	2005	Ministry of Construction (MOC)
Eleventh Five-Year Plan	2006	NDRC
The State Council Decision on Strengthening Energy Conservation	2006	State Council
Revised Consumption Tax for Larger, Energy-Inefficient Vehicles	2006	MOF and the State Administration of Taxation
Reduced Export Tax Rebates for Many Low-Value-Added But High Energy-Consuming Products	2006	MOF
Top-1000 Energy-Consuming Enterprise Program	2006	NDRC
"Green Purchasing" Program	2006	Ministry of Environmental Protection (MEP) and MOF
Revision of Energy Conservation Law	2007	National People's Congress and NDRC
Allocation of Funding on Energy Efficiency and Pollution Abatement	2007	MOF and NDRC
China Energy Technology Policy Outline 2006	2007	NDRC and the Ministry of Science and Technology
Government Procurement Program	2007	NDRC and MOF
National Phase III Vehicle Emission Standards	2007	
Interim Administrative Method for Incentive Funds for Heating and Metering and Energy Efficiency Retrofit for Existing Residential Buildings in China's Northern Heating Area	2007	MOF
Law on Corporate Income Tax (preferential tax treatment for investment in energy-saving and environmentally-friendly projects and equipment)	2008	NDRC
Allocation of Funding on Energy Efficiency and Pollution Abatement	2008	MOF and NDRC
Appliance Standards and Labeling	Various Years	General Administration of Quality Supervision, Inspection and Quarantine

During the summer of 2006, all participating enterprises signed energy-conservation agreements with local governments committing themselves to reaching the energy-savings target by 2010. In addition, the energy-saving target has been added to the provincial government cadre-evaluation system. Preliminary data indicate that the large majority of Top1,000 enterprises are meeting their interim targets (Price et al., 2009).

Government Funding for Private Investment

In 2007, the Chinese government allocated 23.5B RMBY (about $3 billion at that time) to projects for improving energy efficiency and reducing pollution (MOF, 2008). This funding supported the launch of the Ten Key Projects (described above), the elimination of inefficient facilities, and the installation of measures to protect the environment. These funds are also being used to award 200 to 250 RMBY ($26 to $33) for every tonne of coal equivalent an enterprise saves through the implementation of five of the Ten Key Projects in energy-intensive industries, coal-fired industrial boilers, district heating using cogeneration systems, and buildings (Jiang; 2006; Lu 2006).

In 2008, the total allocation for energy conservation, emissions reduction, and ecological improvement was doubled to 42B RMBY (about $6 billion) (MOF, 2008). This funding includes 7.5B RMBY ($1 billion) for awards for the Ten Key Projects and 4B RMBY ($0.6 billion) for phasing out inefficient industrial plants.

Overall Results

In 2006, the energy intensity of the Chinese economy decreased by 1.7 percent, the first decrease in this measure since 2001 (Zhou et al., 2009). Although this was a significant achievement, the reduction was well below the trajectory needed to achieve the goal of a 20- percent reduction in energy intensity by 2010. In 2007, however, energy intensity declined by 3.7 percent, and in 2008, it was reduced by 4.6 percent (Zhou et al., 2009). In the first quarter of 2009, preliminary data indicate an even greater reduction (China View, 2009).

Although the impact of the world economic crisis on energy intensity in China is difficult to predict, it now appears that China is likely to meet its 20-percent energy-intensity reduction target for 2010. Such savings represents a decrease of 1.5 billion metric tons of CO_2 (Lin et al., 2007), a very large number by any measure.

REFERENCES

Andrews-Speed, P. 2009. China's on-going energy efficiency drive: origins, progress, and prospects. *Energy Policy,* 37(4): 1331–1344.

China View. 2009. China's energy intensity down 2.9% in Q1: statistics office. Available online at *http://news.xinhuanet.com/english/2009-04/29/content.* April 29.

EPA (Environmental Protection Agency). 2008. Inventory of U.S. Greenhouse Gas Emissions and Sinks: 1990–2006. Washington, D.C.: EPA.

Economy, E.C. 2007. The Great Leap Backward. Foreign Affairs 86(5): 38–59. Economy, E. 2009. *Personal communication,* May 4, 2009.

IEA (International Energy Agency). 2003. Energy Balances of non-OECD Countries. Paris: IEA.

IEA. 2008. World Energy Outlook. Paris: IEA.

Jiang, Y. 2006. Personal communication with Jiang Yun, China Energy Conservation Association, November 6, 2006.

Levine, M., and N. Aden. 2008. Global Carbon Emissions in the Coming Decades: The Case of China. Pp. 19–38 in Annual Review of Environment and Resources, Vol. 33. Palo Alto, Calif.: Annual Reviews.

Lin, J. 2005. Trends in Energy Efficiency Investments in China and the United States. LBNL576921. Berkeley, Calif.: Lawrence Berkeley National Laboratory.

Lin, J., N. Zhou, M. Levine, and D. Fridley. 2007. Taking out 1 billion tons of CO_2: the magic of China's 11th Five-Year Plan? LBNL-62886. Berkeley, Calif.: Lawrence Berkeley National Laboratory.

Lu, Y. 1993. Fueling One Billion: An Insider's View. London, U.K.: Paragon Press.

Lu, X. 2006. Personal communication with Lu Xinming, director, Department of Environment and Resources Conservation, National Development and Reform Commission, September 9, 2006.

MOF (Ministry of Finance). 2008. Ministry of Finance sets aside 27 Billion RMB¥ for supporting energy saving and emission reduction. Available on line at *http://www.mof.gov.cn/caizhengbuzhuzhan/ zhengwuxinxi/ caizhengxinwen/200805/t20080519 _29017.html.*

NBS (National Bureau of Statistics). 2008, and various years. China Statistical Yearbook. Beijing, China: China Statistics Press.

NDRC (National Development and Reform Commission). 2004. Medium and Long-Term Energy Conservation Plan. Beijing: NDRC

Price, L., W. Xuejun, and Y. Jiang. 2009. The challenge of reducing energy consumption of the top 1,000 largest industrial enterprises in China. Energy Policy (special issue). In press. Available online at *http://www.sciencedirect.com/science/*

Sinton, J., and M. Levine. 1998. Energy efficiency in China: accomplishments and challenges. Energy Policy 26(11): 813–829.

USGS (U.S. Geological Survey). 2009. Mineral Commodity Summaries: Cement. Washington, D.C.: USGS.

Zhou, N., M. Levine, and L. Price. 2009. Overview of Current Energy Efficiency Policies in China. Submitted to Energy Policy.

End Notes

[1] Data on investment in energy efficiency prior to 1981 are not available, but investments during these years were undoubtedly much smaller than the 10 percent figure of 1981.

[2] If one assumes a two-year payback on the investments in energy efficiency, then the investment level was sufficient to achieve more than half of the decrease in energy intensity sought by Deng.

[3] This is based on forecasted growth of energy-related CO_2 emissions by IEA (2008).

[4] These statements are based on a large number of interviews by the lead author of this paper with government officials, researchers, industrialists, and academics during the 1990s and early 2000s.

[5] RMBY is the Chinese currency, valued at 6.8RMBY per $U.S.

[6] Many of the energy-supply enterprises were still under the control of the state, either directly as state-owned enterprises or indirectly through the positions of their leaders in government (e.g., the heads of the national oil companies are officially government ministers, even though the companies are mostly privately owned).

[7] Another interesting policy is the responsibility system (also called the “one vote veto”) in which a government official or manager of a state-owned enterprise cannot advance without meeting an energy-intensity target. Thus an individual may meet all of the criteria except the energy-intensity target with very high marks and still flunk the performance evaluation, with significant adverse consequences (Zhou, et al., 2009).

[8] For a detailed review of energy-efficiency policies initiated during the 11th Five-Year Plan and after the announcement of the 20-percent intensity goal, see Lin et al., 2007.

[9] The other six have relatively small impacts.

[10] In 2006, financial support from the government for this purpose was "64M RMBY ("$8 million). The United Nations Development Program/Global Environmental Facility added about "8M RMBY ($1 million) to the total (Jiang, 2006). This is based on a currency conversion of $1 = 8.00419 RMB¥ (average rate of June 2006).

In: Going Green in China: Policy and Potential ISBN: 978-1- 62100-691-6
Editors: Spencer B. Talscott

Chapter 4

CHINA'S COAL: DEMAND, CONSTRAINTS, AND EXTERNALITIES*

Nathaniel Aden, David Fridley and Nina Zheng

DISCLAIMER

This document was prepared as an account of work sponsored by the United States government. While this document is believed to contain correct information, neither the United States government nor any agency thereof, nor The Regents of the University of California, nor any of their employees, makes any warranty, express or implied, or assumes any legal responsibility for the accuracy, completeness, or usefulness of any information, apparatus, product, or process disclosed, or represents that its use would not infringe privately owned rights. Reference herein to any specific commercial product, process, or service by its trade name, trademark, manufacturer, or otherwise, does not necessarily constitute or imply its endorsement, recommendation, or favoring by the United States government or any agency thereof, or The Regents of the University of California. The views and opinions of authors expressed herein do not necessarily state or reflect those of the United States government or any agency thereof, or The Regents of the University of California.

---

* This is an edited, reformatted and augmented version of the Ernest Orlando Lawrence Berkeley National Laboratory publication, LBNL-2334E, dated July 2009.

EXECUTIVE SUMMARY

- China has been, is, and will continue to be a coal-powered economy. In 2007 Chinese coal production contained more energy than total Middle Eastern oil production. The rapid growth of coal demand after 2001 created supply strains and bottlenecks that raise questions about sustainability.
- Urbanization, heavy industrial growth, and increasing per-capita income are the primary interrelated drivers of rising coal usage.
  - In 2007, the power sector, iron and steel, and cement production accounted for 66% of coal consumption.
  - Power generation is becoming more efficient, but even extensive roll-out of the highest efficiency units would save only 14% of projected 2025 coal demand for the power sector.
  - A new wedge of future coal consumption is likely to come from the burgeoning coal-liquefaction and chemicals industries. If coal to chemicals capacity reaches 70 million tonnes and coal-to-liquids capacity reaches 60 million tonnes, coal feedstock requirements would add an additional 450 million tonnes by 2025.
  - Even with more efficient growth among these drivers, China's annual coal demand is expected to reach 3.9 to 4.3 billion tonnes by 2025.
- Central government support for nuclear and renewable energy has not reversed China's growing dependence on coal for primary energy. Substitution is a matter of scale: offsetting one year of recent coal demand growth of 200 million tonnes would require 107 billion cubic meters of natural gas (compared to 2007 growth of 13 BCM), 48 GW of nuclear (compared to 2007 growth of 2 GW), or 86 GW of hydropower capacity (compared to 2007 growth of 16 GW).
- Ongoing dependence on coal reduces China's ability to mitigate carbon dioxide emissions growth. If coal demand remains on a high growth path, carbon dioxide emissions from coal combustion alone would exceed total US energy-related carbon emissions by 2010.

- Within China's coal-dominated energy system, domestic transportation has emerged as the largest bottleneck for coal industry growth and is likely to remain a constraint to further expansion. China has a low proportion of high-quality reserves, but is producing its best coal first. Declining quality will further strain production and transport capacity. Furthermore, transporting coal to users has overloaded the train system and dramatically increased truck use, raising transportation oil demand.
- Growing international imports have helped to offset domestic transport bottlenecks. In the long term, import demand is likely to exceed 200 million tonnes by 2025, significantly impacting regional markets.

I. Overview

This study analyzes China's coal industry by focusing on four related areas. First, data are reviewed to identify the major drivers of historical and future coal demand. Second, resource constraints and transport bottlenecks are analyzed to evaluate demand and growth scenarios. The third area assesses the physical requirements of substituting coal demand growth with other primary energy forms. Finally, the study examines the carbon- and environmental implications of China's past and future coal consumption.

The following three sections address these areas by identifying particular characteristics of China's coal industry, quantifying factors driving demand, and analyzing supply scenarios. Section two reviews the range of Chinese and international estimates of remaining coal reserves and resources as well as key characteristics of China's coal industry including historical production, resource requirements, and prices. Section three quantifies the largest drivers of coal usage to produce a bottom-up reference projection of 2025 coal demand. Section four analyzes coal supply constraints, substitution options, and environmental externalities. Finally, the last section presents conclusions on the role of coal in China's ongoing energy and economic development.

II. China Coal Resources and Production

It is widely agreed that China possesses the third largest coal resources in the world, behind the United States and Russia.[1] However, there is no such consensus on the precise extent or availability of China's coal resources. Estimates vary among Chinese and international sources, in reference to differing types of coal, and due to confusion of reserve and resource categories.

2.1. Reserve Estimates

China's reserve estimates have declined significantly from early 20th-century estimates of as high as 1 trillion tonnes in the 1920s. Reserves estimates fell from 700 billion tonnes in the 1950s to 300 billion tonnes in the 1970s. After a series of extensive national resource surveys, China reported their reserves to the World Energy Council (WEC) as 114.5 billion tonnes in 1992. The WEC has continued to use this number in spite of more than 25 billion tonnes of Chinese domestic coal consumption in subsequent years. The implication of the WEC's frozen reserve estimate is that resources have been converted into reserves at a rate equal to annual consumption—or that the central government has not formally reported revised numbers. The WEC reserve estimate is commonly cited in other references, including the annual BP Statistical Review of World Energy. In 2003, based on a new national survey of resources, China's Ministry of Land and Resources released an updated estimate of 189 billion tonnes total coal reserves.[2] Figure 1 displays China's coal resource and reserve estimates, from 189 billion tonnes of indicated economically-viable coal, 326 billion tonnes of reserve base, on to a 1,021 billion-tonne estimate of total resources.[3] The reserve base encompasses all identified resources that meet the physical and economic criteria for extraction as well as those that have potential for becoming extractable in a time horizon beyond proven technology and current economic conditions.

According to the National Bureau of Statistics (NBS), coal completely dominates China's fossil fuel reserve base. Figure 2 illustrates the energy content of the reserve base (基础储量; translated in Chinese sources as "ensured reserves") of coal, oil and natural gas published in the 2008 China Statistical Yearbook. Here NBS reports a reserve base of 2.8 billion tonnes of petroleum and 3 trillion cubic meters of natural gas, along with the coal reserve base of 326 billion tonnes raw coal. For China, coal constitutes 97% of the fossil fuel reserve base by energy content; in contrast, in 2007, the WEC estimated that 62% of world conventional fossil fuel reserves were coal by energy content; 19% were oil, and 19% natural gas.

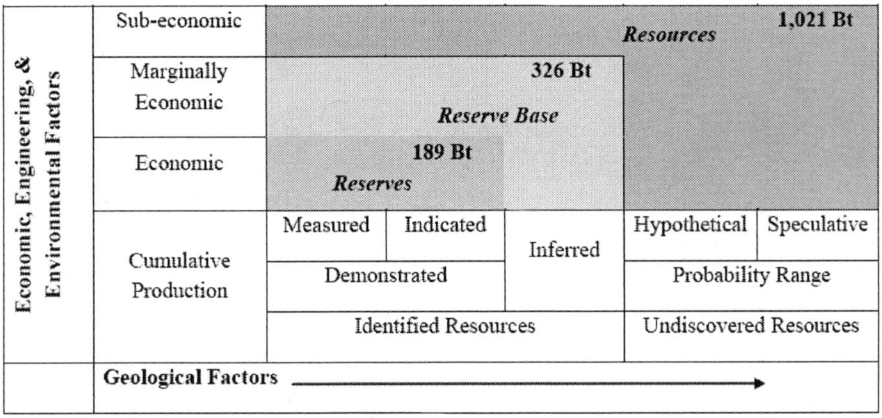

Figure 1. China Coal Resources and Reserves, 2008[4].

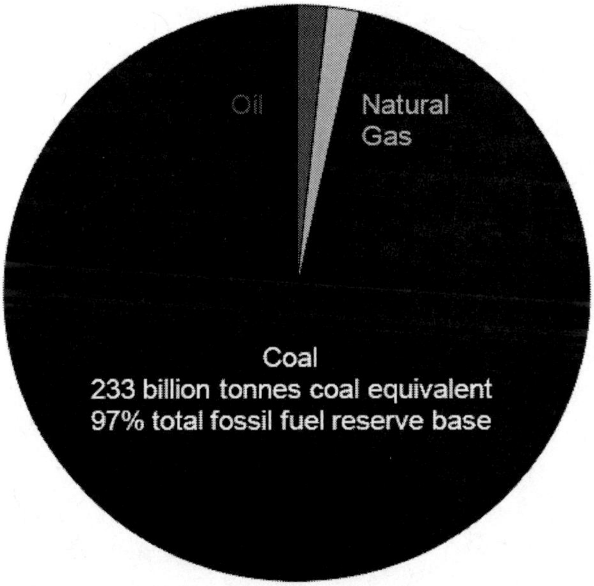

Figure 2. Energy Content of China's Fossil Fuel Reserve Base, 2007.

2.2. Reserves by Location, Quality, and Depth

Most of China's coal reserves are concentrated in the northern provinces of Shanxi and Inner Mongolia. Southern coastal areas with the highest GDP growth, such as Guangdong and Fujian, are among China's least coal-abundant provinces. Map 1 illustrates the provincial distribution of indicated economic reserves according to the Ministry of Land and Resources 2003 Survey.

Three provinces were estimated to have indicated economic coal reserves of at least 15 billion tonnes. Shanxi province has the richest coal endowment in China—almost 59 billion tonnes. Inner Mongolia trails Shanxi with an estimated 47 billion tonnes. Shanxi's western neighbor province, Shaanxi comes in a distant third place with an estimated 16 billion tonnes of indicated economic coal reserves.

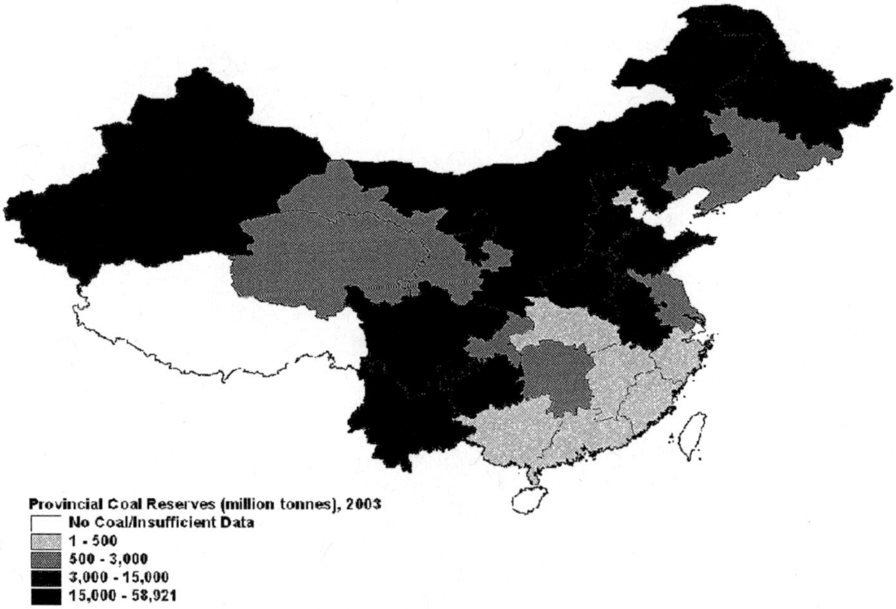

Map 1. China Provincial Coal Reserves (Mt), 2003.

The quality of China's coal production ranges from 3,446 kcal/kg in Guangxi to 6,245 kcal/kg for Ningxia coal.[5] China's highest quality coal reserves are concentrated in the four provinces of Gansu, Ningxia, Shaanxi, and Shanxi. On a national level, 54% of China's coal reserves are classified as bituminous coal by volume, versus 29% sub-bituminous and 16% lignite.[6] Shanxi province in Northern China accounts for 31% of China's indicated economic reserves. In addition to being the most plentiful, Shanxi coal is among the country's highest quality—average provincial calorific values are 6,242 kcal/kg, versus the national average of 5,350 kcal/kg. The average heating value for American coal is 5,600 kcal/kg.[7]

The average depth of China's coal mines is 456 meters. Whereas northern China has the most abundant and highest quality coal, Xinjiang province (in far western China) has more than half of coal reserves located less than 1,000 meters below the surface. Only 27% of northern Chinese coal is located less than 1,000 meters below the surface, compared to 40% of total Chinese coal.[8] Mines in eastern China are particularly deep, with an average depth of 600 meters. Although the average sulfur content of Chinese coal ranges up to 5%, it increases with depth in north China, suggesting that sulfur content will rise over time.[9] Due to deep coal resources, more than 90% of Chinese mines are pithead mines, compared with less than 40% in the United States, Australia, and India. The dominance of pithead mines in China lowers the coal recovery rate and increases extraction costs relative to open-pit mining.

2.3. Production by Coal Type

Since 1980, China maximized its high-quality coal production by primarily focusing on extraction of bituminous coal, with lower production growth of anthracite, lignite and brown coal.

While bituminous coal is estimated to comprise 54% of China's coal reserves, the bituminous share of total production varied from 73% to 78% between 1980 and 2006.[10] Figure 3 illustrates the growth of bituminous, anthracite, and lignite production.

As illustrated in the figure above, China's production of bituminous coal has risen steadily, with comparable increases in both coking-grade bituminous coking and lesser-quality bituminous coal.

As the largest share of total coal production, bituminous coal consisted of 76% of China's total coal production in 2006. With the exception of the late 1990s contraction, bituminous coal production has increased each year, rising from 467 million tonnes in 1980 to 1,785 million tonnes in 2006. During this period, an average of 47% of China's total coal production was composed of coking-quality coal while 30% composed of other general bituminous coal. There were no major fluctuations in the relative proportions of coking and other general bituminous coal until 2004 when the coking-quality share of total production began to diminish.

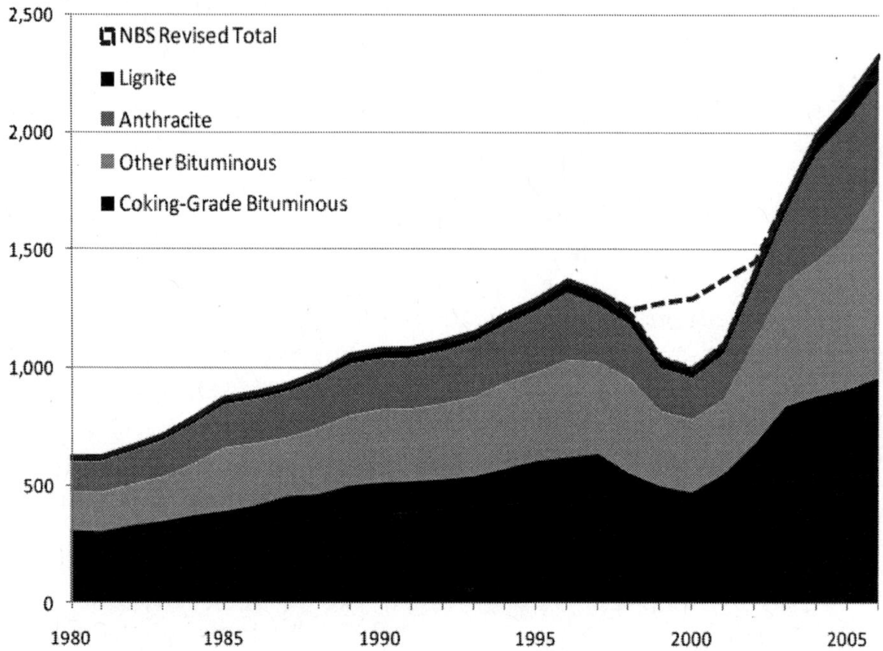

Note: production data are reported by coal quality rather than end use; 1998-2002 production data from China Coal Industry Yearbook have not been revised to accord with revised aggregate production numbers (as shown by the dotted line, for which coal quality is unknown).

Figure 3. Raw Coal Production by Type of Coal, 1980-2006.

Anthracite coal production follows bituminous coal production with an approximately 20% share of total production throughout the same time period. Its production volume increased nearly four-fold from 129 million tonnes in 1980 to 442 million tonnes in 2006. This was followed by the production of lignite and brown coal, which also increased four-fold in production volume from 24 to 105 million tonnes. Whereas lignite and brown coal accounted for 4% of total coal production in 2006, China's coal reserves are estimated to be 16% lignite by volume.[11]

2.4. Production by Mine Ownership

Unlike the relatively steady trends in the production shares of coal types, the shares of different types of coal production ownership in China has changed drastically since 1980. Overall, the production share of centrally administered state-owned mines has decreased, from a share of 56% in 1980 to 47% in 2005, while the shares of collective and private mines have both increased. In parallel with centrally-administered mines, shares of local state-owned mines have decreased, falling from its highest share of 26% in 1980 to its lowest share of 14% in 2005. Among different types of local state-owned mines, provincial mine ownership has decreased the most, followed by prefectural and county ownership. Government policy influences the collective and private mines' share of production through regulation of mine safety, production volume, and coal quality. The ongoing lack of coordination among smaller private mines accounts for China's relatively low resource extraction rates.

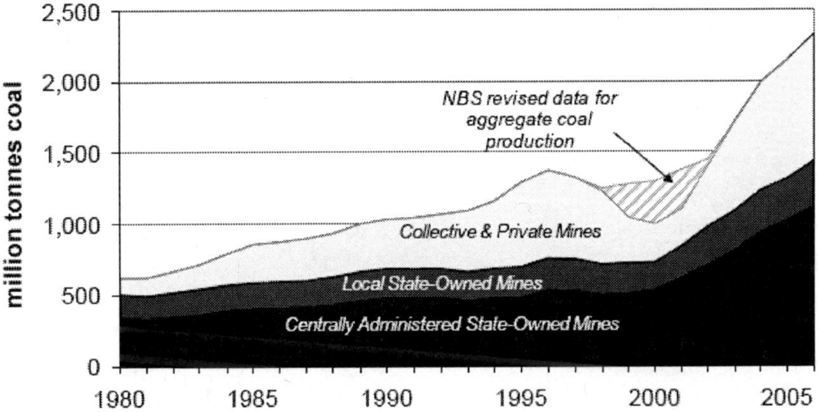

Source: Coal Industry Yearbook, various years; China Statistical Yearbook, various years.

Note: mine ownership data are from the Coal Industry Yearbook, which has not published any revisions; additional production is likely to have come from collective and private mines.

Figure 4. Coal Production by Mine Ownership, 1980-2006.

In terms of output production growth, the average annual growth between 1980 and 1995 was low for centrally administered state-owned mines and even lower for local state-owned mines. Local state-owned mines' coal production, in particular, often underwent negative annual growth. In contrast, average production growth for collective and private mines was very high before 1996 and almost all of the gains in national coal production can be attributed to these mines. The growth in production output from collective and private mines during this period resulted largely from many households responding to the government's encouragement for small mines.

China's 11th Five-Year Plan presented targets for coal industry restructuring and reform by 2010. Most of these targets are oriented towards industry consolidation and increased efficiency. Lack of technology among small Township and Village Coal Mines (TVCM) partially explains their very low average extraction rate of approximately 15%.[12] NDRC-mandated mine closures were expected to reduce production by 200 million tonnes—perhaps a reason that mines were re-opened in 2008 as national supply tightened. Industry consolidation and mechanization are central efforts in the NDRC's target of raising average recovery rates from the 2005 official national average of 46% to 50% by 2010.[13] Closure of smaller mines is a key component of the Eleventh Five Year Plan. In 2006, the National Development Reform Commission (NDRC) announced that only mines producing at least 300,000 tonnes per year would be considered for approval; the NDRC has also announced that it intends to reduce the number of small mines to 10,000 by 2010.[14] However, local officials are more severely punished for mining accidents and safety violations than low recovery rates.[15] This dynamic has limited the government's control of small mines and highlighted tradeoffs between improved aggregate coal recovery rates and local-scale investment and regulatory costs.

2.5. Water Resources and Requirements

China's average per-capita water availability is 2,300 m^3, which is approximately one-third world average. While coal reserves and production are concentrated in the north, limited water resources have posed problems for increased production and coal washing. North China contains the vast majority of domestic coal reserves and more than half the population; however, the region has access to only 20% of national water resources. Average per-capita water availability in North China is 271 m^3—one eighth the national average and $1/25^{th}$ the world average.[16] As a result of imbalanced rainfall distribution, North China has an arid climate where its naturally dry conditions were worsened by repeated droughts in the 1990s. The water shortage problem has been exacerbated with increased coal mining as well as increased demand from a growing population in the north. In contrast, southern China has ample water resources and often faces the threat of annual flooding.

As for China's mining industry, it has been estimated that the extraction of 1 tonne of coal requires 53 – 120 liters of water, depending on the location and depth of the coal.[17] An additional 4 tonnes of water is needed for coal washing, which can reduce sulfur and particulate content while increasing energy content of raw coal. As a result of high water requirements, the portion of washed coal production has remained low. Finally, water is also needed for cooling, beneficiation, and other plant operations within thermal electric power plants. The unit water requirement of coal-fired electricity generation is estimated at 0.164 m^2/GJ, compared to 0.109 m^2/GJ for natural gas fired electricity generation.[18]

2.6. International Coal Prices and Trade

In parallel with the surge of energy consumption after 2001, domestic Chinese coal prices moved from stable levels under $40 per tonne to more than $70 per tonne for coking coal in 2006. From 1998 to 2002, domestic prices for steam and coking coal varied between $26 and $38 per tonne. As illustrated in Figure 5, domestic steam and coking prices have risen unsteadily since 2002, with steam prices converging with coke prices three times before resuming lower levels.

Overall, domestic coal prices increased when electricity demand and power plant construction soared. Prices reached historical highs in 2008. At Qinhuangdao, China's main coal pricing terminal, coking coal prices have risen to $137/tonne in 2008, and steam coal prices have strengthened as well.[19] Datong premium blend (6,000 kcal/kg) traded in May 2008 at 655-670 RMB per tonne FOB ($94-96/tonne).[20] By early 2009, prices dropped back in the face of economic slowdown, bringing Datong premium blend back to 590 RMB per tonne ($86/tonne)—still more than double prices earlier in the decade.[21]

From an international perspective, China's domestic coal prices after 2000 have been increasing more consistently than other countries, such as Australia. In contrast to Chinese domestic prices, Australia's FOB export value for coking coal was relatively stable before 2004, as was the value for its steam coal before 2003. After 2004, however, Australia's FOB export value for coking coal significantly increased by $40 per tonne while its steam coal export value also increased by $20 per tonne. The two price trends have become increasingly interrelated: as China's coal exports have declined and imports increased because of a tighter domestic market, import demand from China's former customers has shifted to other Asia-Pacific countries as has China's import demand, pushing up benchmark prices in Australia.

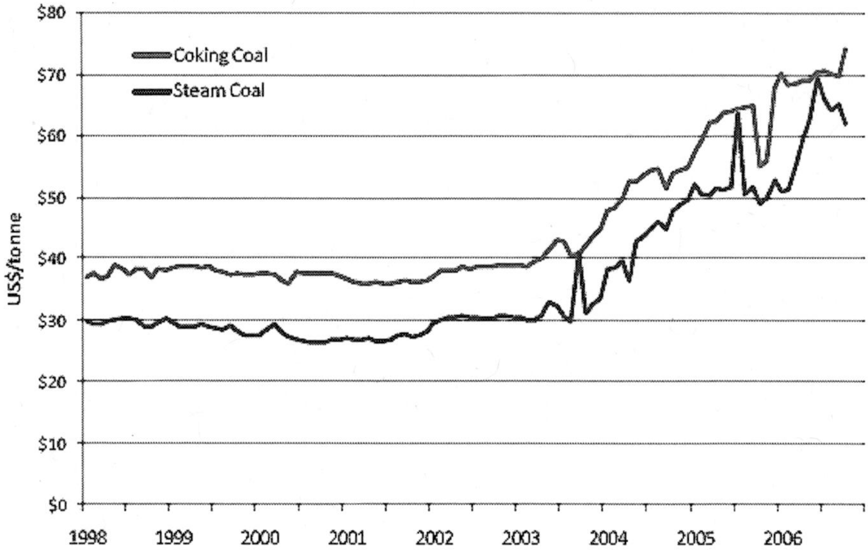

Data Source: Beijing Energy Efficiency Center, available at: http://www.beconchina.org/energy price.htm.

Figure 5. Nominal Domestic Prices for Steam and Coking Coal, 1998 – 2006.

III. CHINESE COAL DEMAND

Although China has a smaller coal resource endowment than the United States and Russia, it is more dependent on coal as a primary source of energy.[22] Coal's dominant share of primary energy ultimately derives from its 97% share of total fossil fuel endowments (Figure 2). Figure 6 shows that coal comprised 70 percent of China's 2006 total primary energy consumption, compared to 24 percent in the United States and 16 percent in Russia.

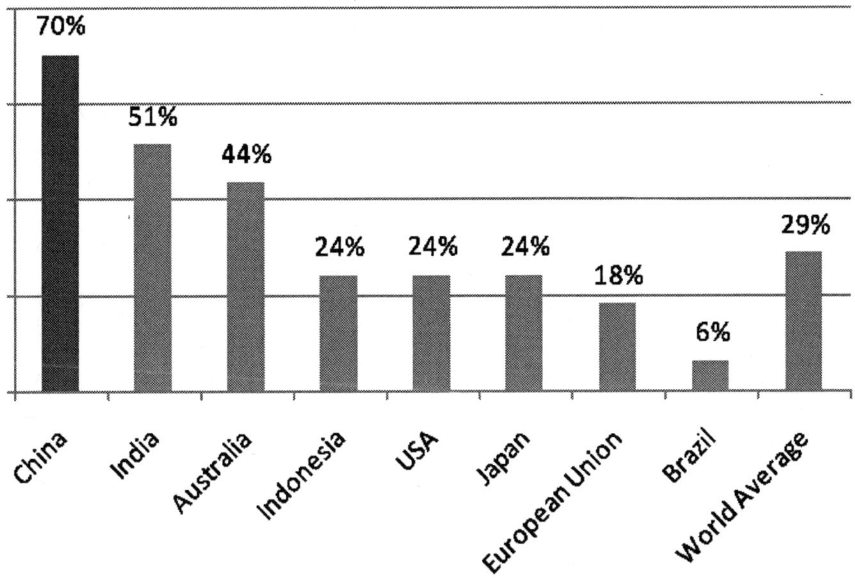

Source: BP Statistical Review of World Energy 2008.

Figure 6. Comparative Coal Dependence of Primary Energy Consumption, 2007.

The high level of domestic coal dependency in China is partly a result of the high degree of resource self-sufficiency (Figure 7). Given China's unbalanced fossil resource distribution, self sufficiency and energy security lead to high levels of coal usage. Although countries such as Indonesia and the US are completely self-sufficient in coal supply and are net exporters, coal is a much smaller percentage of their total primary energy consumption because of the availability of other fossil fuel resources in large volumes.

While China's dependence on coal as a primary energy source declined between 1980 and 2000, coal's share rebounded from 68% in 2000 to 70% in 2006. In the face of oil and natural gas price rises, coal's share may be further enhanced as coal-based fuel and feedstock substitution programs are implemented. In this sense, China's coal crunch is not merely demand-derived, but is part of a larger energy crisis rooted in supply concerns in the oil and natural gas industries.

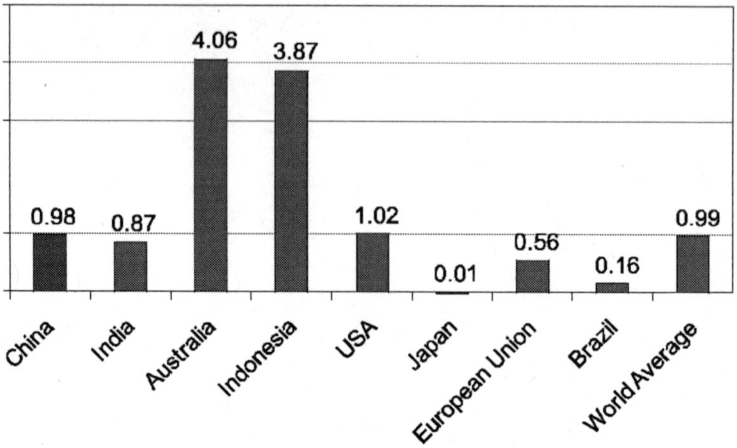

Source: BP Statistical Review of World Energy 2008.

Figure 7. Ratio of Domestic Coal Production to Consumption, 2007.

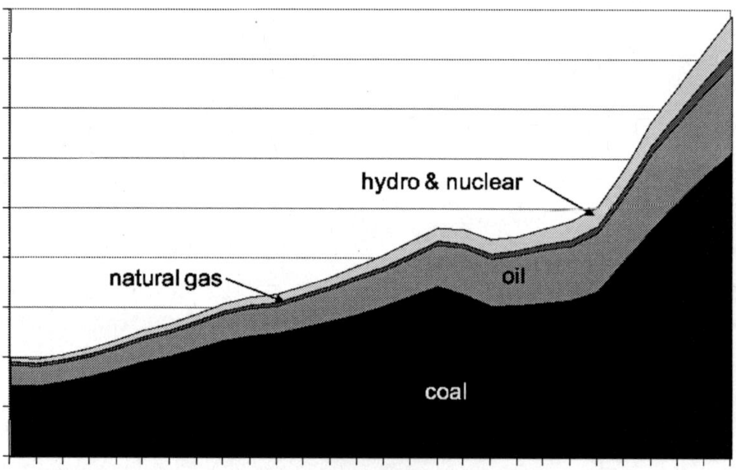

Figure 8. China Primary Energy Consumption, 1980-2007.

3.1. Actual and Forecasted Coal Use

Since 2003 actual Chinese coal consumption has exceeded forecasts from the IEA and the EIA, as well as from China's own National Development and Reform Commission (NDRC). As shown in Figure 8, the average annual growth rate of coal consumption has exceeded 10% since 2001. China's reported 2007 coal consumption of 2,580 million tonnes exceeded the IEA's WEO 2000 forecast for 2020 coal consumption of 2,473 million tonnes. In its 11th Five-Year Plan for Coal Industry Development, the NDRC forecast that China's coal production would reach 2.6 billion tonnes in 2010—a level that was exceeded in 2008. In their 2008 forecasts, the IEA and US EIA estimate China 2025 coal consumption at 4.7 and 4.6 billion tonnes, respectively (Figure 9).

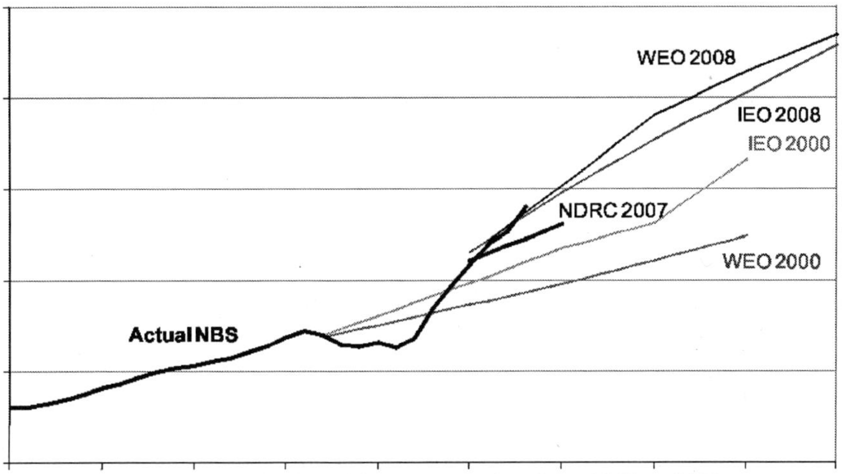

Note: Actual NBS data are displayed for consumption from 1980 to 2008.

Figure 9. Historical and Forecast Chinese Coal Consumption, 1980-2025.

3.2. Drivers of Demand: Urbanization, Heavy Industry, and Rising Income

The long-term trend in China's coal usage is a monotonic shift from direct end use to transformation, primarily through thermal electricity generation. Between 2000 and 2006, total direct end use of coal dropped from 35% to 26% of annual coal consumption. Over the same period, power generation increased from 42% to 50% of the total. Industry end use of coal increased on an absolute basis, but declined from 26% to 20% of total consumption. The shift from end use to transformation of coal is driven by inter-related processes of urbanization, heavy industry growth, and rising per-capita consumption.

Between 1990 and 2007 the urban share of China's population surged from 26% to 45% of total. The addition of 290 million urban residents had three direct effects on China's energy system: cement and steel had to be produced for building and infrastructure construction, electricity demand rose with home appliance ownership, and overall demand was stimulated by the replacement of rural non-commercial (largely biomass) energy with urban commercial energy services—principally electricity. This study examines the coal demand implications of 66% urbanization in 2025.

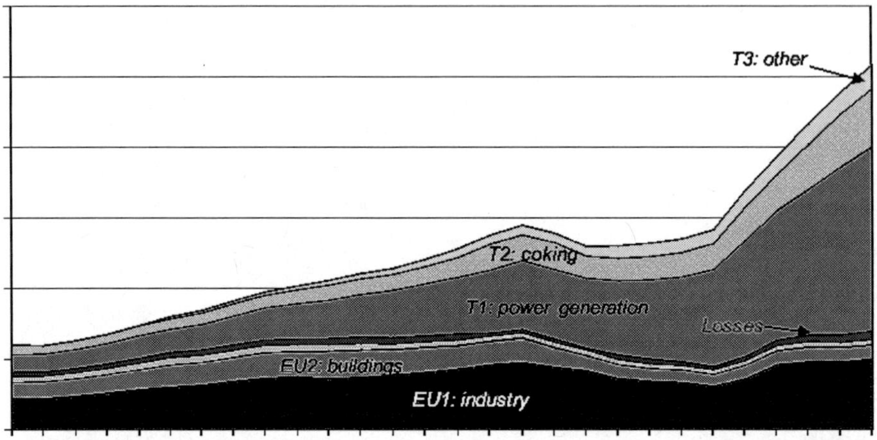

Source: NBS, 2007. EU2 = commerce + other (government) + residential; T3 = heating + gas production.

Figure 10. Chinese Coal Consumption by End Use and Transformation, 1980-2007.

As shown in Figure 10, the surge of coal consumption between 2000 and 2007 was largely driven by the rapid rise in electricity demand. Coal-fired power generation accounted for 56% of the marginal increase in coal use between 2000 and 2005 (Figure 11). Over the same period, growth in coal use for power generation was followed by the growth in coal use for coke production (18%), the end-use of coal for production of building materials (6%), delivered heating (district heating) (6%), and chemicals production (3%) as the largest growth drivers.

Under the assumption that China will maintain its planned growth trajectory, both Chinese and international assessments indicate that urbanization will continue to be a primary driver of energy consumption. In its 11th Five-Year Plan, the NDRC forecast China's population will expand from 1.31 billion in 2006 to 1.36 billion in 2010, of which 47% are expected to be urban residents. By 2025, the UN expects 57% urbanization and the McKinsey Global Institute forecast 66%, out of a total population of 1.403 billion.[23] China's urbanization is characterized by a combination of economically-motivated rural-urban migration and establishment of new urban centers in formerly rural areas. The requirements of urbanization—infrastructure, housing, services, transportation, and a shift to commercial energy use by residents—drive growth in the heavy industries that supply the materials, products and energy for urban construction and residential services.

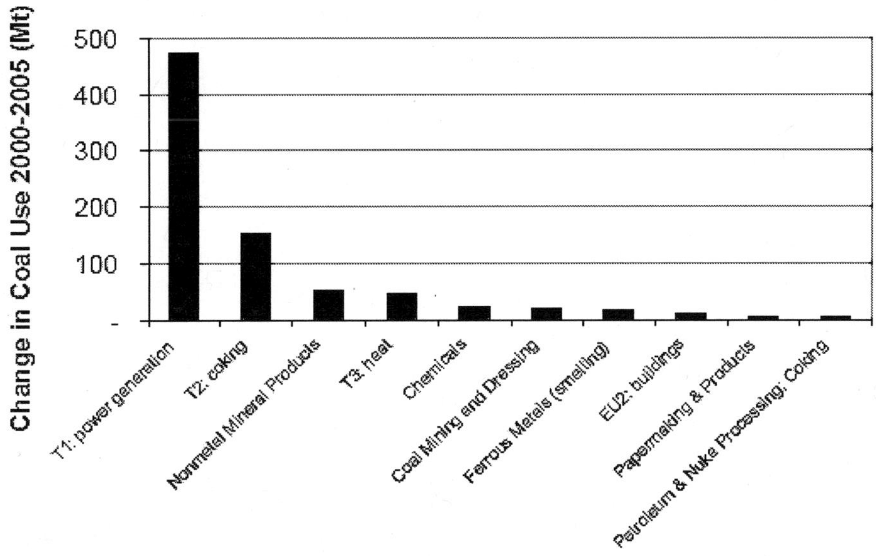

Figure 11. Drivers of Growing Chinese Coal Use, 2000-2005.

3.3. Power generation

Coal increasingly dominates China's electricity generation system. Since 1980 coal's share of electricity generation capacity has grown steadily, from 69% to 78% in 2007. As shown in Figure 12, the absolute amount of coal-fired capacity grew at an average annual growth rate of more than 12% between 2000 and 2007, from 238 to 554 GW. In spite of this rapid growth, China's per-capita electricity generation capacity is comparatively low (0.5 kilowatts per person in 2007). Japan's per capita generation capacity, for example, was 1.9 kw per person in 2007.

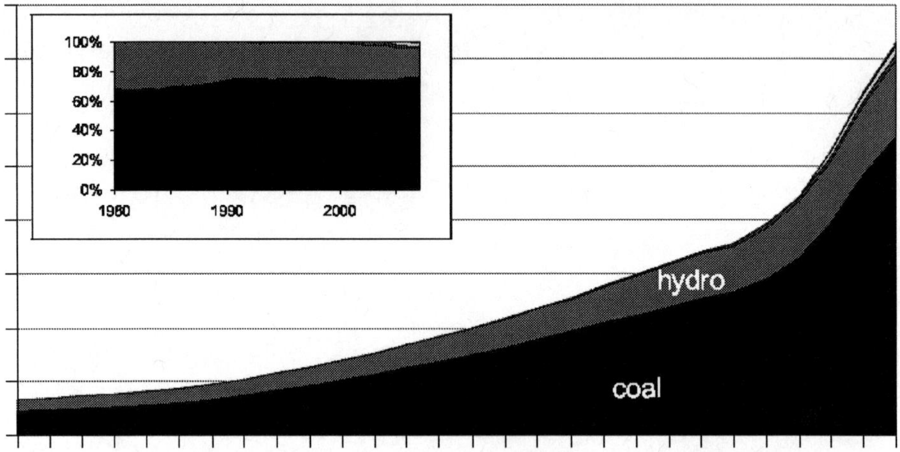

Note: nuclear, natural gas, and renewable-based electricity capacity displayed above hydropower in this figure.

Figure 12. Electricity Generation Capacity by Fuel, 1980-2007.

3.3.1. Electricity Pricing

In the beginning of 2008, 42% of China's statistically-sampled thermal power plants lost money due to the convergence of government caps on retail electricity prices, increased repair and maintenance costs related to severe snowstorms, and rapidly rising coal prices.[24] While electricity pricing reform policies were initiated in the last decade, prices are still set and controlled at the central, provincial and even municipal and county levels. Specifically, power producers' prices for selling electricity to the provincial power companies and to the State Power 98 Corporation must be reviewed and approved by provincial bureaus and the State Development and Planning Commission and the State Pricing Bureau, respectively .[25] In 2004, the National Electricity Regulatory Commission established a policy stipulating the adjustment of electricity prices if average coal prices fluctuate more than 5% withi n 6 months.[26] In practice, however, Hydo electricity prices were not allowed to rise in spite of rapidly increasing coal prices in 2007. To avoid exacerbating inflation and its social and economic impacts, a cap on electricity prices was implemented to control the consumer price index.

In conjunction with national government price caps, provincial governments adjust electricity tariffs among end-use sectors. Retail electricity prices are set in provincial catalogues for eight categories of users, including the introduction of a "commercial" category in the 1990s for profitable enterprises in the commercial sector. While relative tariff levels vary by region, a prevailing trend is relatively low catalogue tariffs for residential, industrial, chemical, agricultural and irrigation users and higher tariffs for non-residential lighting and commercial users.[27] Additionally, time-of-day tariffs also vary between different user classes. For example, residential and irrigation users are exempt from the time-of-day tariffs while heavy industry, chemical plants and agriculture users face significantly lower tariffs.

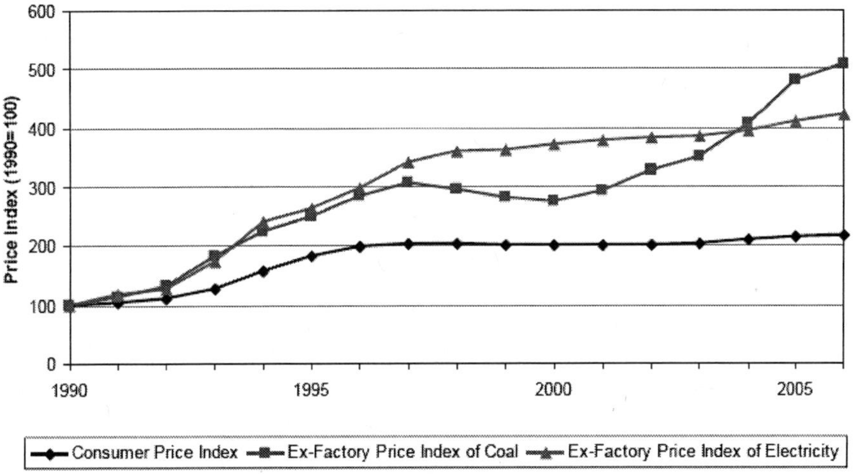

Source: NBS, 2007.

Figure 13. Price Indices of Coal and Electricity, 1990-2006.

The disparity between controlled electricity prices and more market-oriented coal prices has created cost incentives for generators to maintain low coal inventories, thereby rendering them vulnerable to supply disruptions, as illustrated in extensive power outages during the spring festival holiday in 2008. Controlled electricity prices have also provided a profit incentive for coal producers to export rather than taking a lower power plant price on the domestic market. In June 2008, the government acknowledged that the disparity between coal and electricity prices was unsustainable and increased tariffs modestly by ¥0.025 per kWh. This increase may not have covered the increased coal costs to generators, but traditionally such price changes have been undertaken in "stairstep" fashion over time in an attempt to minimize the inflationary impact of the adjustments—i.e., further adjustments are likely. Sustained high prices both domestically and internationally may stimulate a return to Chinese domestic coal price controls in order to stem electricity generators' financial losses if inflation is considered too serious to allow large increases in electricity pricing.

3.3.2. Thermal Technology: Efficiencies and Costs

The average efficiency of thermal electricity generation is comparatively low in China due to the prevalence of small, outdated coal-fired power plants. According to the China Electricity Council, average capacity of coal-fired power plants was 58 MW in 2006, and only 45% of plants had a capacity of at least 300 MW. Larger-scale plants are more capital intensive but require less coal per unit of output—600 MW plants, for example, are on average 17% more energy efficient than 100 MW plants.[28] Thermal coal electricity generation efficiency also varies by plant technology type. This study examines four coal-thermal generation technology types currently or planned to be deployed in China; Table 1 shows the average heat rate and efficiency for each type.

Table 1. Average Heat Rates & Efficiencies for New Coal Thermal Power[29]

Technology	Sub-critical	Supercritical	Ultra-supercritical	IGCC
Heat rate (gce/kWh)	324	300	256	223
Efficiency	38%	41%	48%	55%

Note: IGCC refers to integrated gasification combined cycle technology.

The vast majority of China's thermal power generators use sub-critical combustion technology. Supercritical and ultra-supercritical technology attains higher fuel efficiency by operating at high temperatures and pressures where the boundary between water's liquid and vapor states disappears. According to the World Coal Institute, there were more than 240 supercritical units worldwide in 2006, of which 22 were operated in China.[30] China added 18 GW of supercritical capacity in 2006, bringing the supercritical share of total coal capacity to 6%, up from the China Electricity Council's previous estimate of 4.3% in 2006.[31] The China Huaneng Group began commercial operation of China's first commercial ultra-supercritical plant in November 2006. In 2007, domestic Chinese manufacturers were reported to have orders for 30 new units of ultra-supercritical power generation equipment.[32]

Four scenarios were developed to quantify the effect of new technology adoption on average power plant fleet efficiency. In all of the scenarios total electricity generation capacity grows to 1,576 GW in 2025, of which 1,060 GW are coal-thermal. The scenarios also include 10 GW per year of retired coal-fired capacity between 2009 and 2015, and 5 GW per year thereafter. The efficiency of the retired capacity is estimated at the average fleet heat rate from fifteen years prior to retirement.

The baseline scenario simulates current trends in thermal electricity generation efficiency improvements. Between 2008 and 2015, the baseline scenario assumes 40% of new capacity will use supercritical technology, one gigawatt of ultra-supercritical capacity will be added per year, and the remaining new build will use sub-critical technology. From 2016 to 2025, the baseline scenario assumes 60% of new capacity will use supercritical technology with five gigawatts per year of ultra-supercritical capacity, with the remaining build using sub-critical technology. Average fleet efficiency in the baseline scenario improves from 357 kilograms coal equivalent per kilowatt-hour (kgce/kWh) in 2007 to 323 kgce/kWh (38% efficiency) in 2025.[33] The second scenario assumes that half of all new capacity will use supercritical technology between 2008 and 2025 and the other half will use ultra-supercritical technology. The third scenario assumes that all new build between 2008 and 2025 will use ultra-supercritical technology. The fourth scenario, IGCC, is a hypothetical, most technologically-aggressive scenario—it assumes all build between 2008 and 2010 will use ultra-supercritical technology and all plants thereafter will use IGCC technology.

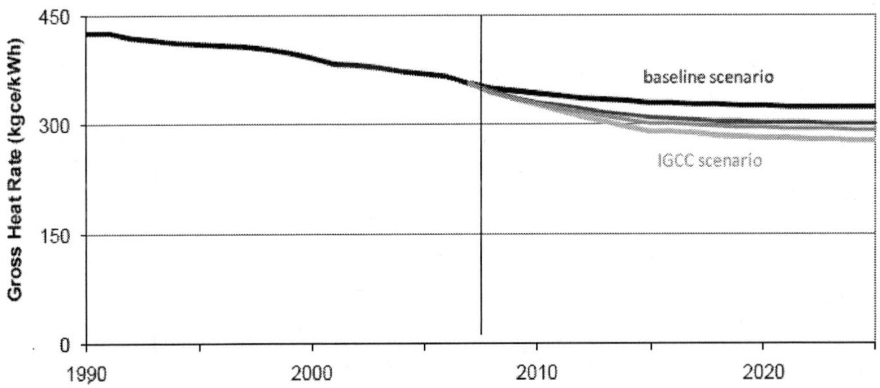

Figure 14. Historical and Hypothetical Efficiency of China's Coal-Fired Electricity Generation, 1990-2025.

Figure 14 shows that the most aggressive adoption of the highest-efficiency coal combustion technology will only improve total fleet efficiency by 14%, from 323 kgce/kWh to 278 kgce/kWh in 2025. Aside from their different levels of thermal efficiency, each of the scenario technologies has varying costs and requirements. In comparison with regular pulverized coal combustion, Chinese researchers found that IGCC plants offer higher input and output flexibility, lower cost CCS possibilities, and require less water—all at higher capital cost and lower reliability.[34] At a coal price of 300 RMB/tonne, Chinese researchers estimate the cost of electricity for an ultra-supercritical plant at 280 RMB/MWh (4 cents per kWh), versus 350 RMB/MWh (5 cents per kWh) for IGCC technology.[35] Without government regulation or financial support, IGCC may have difficulty competing with ultra-supercritical or—if it develops beyond lab-scale—oxy-fuel technology.

3.3.3. Carbon Capture and Storage: Efficiency Penalties and Emissions Mitigation

Carbon capture and storage (CCS) refers to a range of technologies being developed to enable carbon dioxide from fossil fuel combustion to be sequestered in geological sites rather than being emitted to the atmosphere. Carbon dioxide is an innate product of fossil fuel combustion; however, there is a growing consensus that carbon emissions need to be mitigated. The power requirement to operate CCS results in an effective reduction of power plant efficiency. Model-based estimates for the efficiency penalty of CCS technology vary from 14% to 28%.[36] This study uses a 25% efficiency penalty to examine the implications of CCS technology implementation. In addition to its efficiency penalty, published research estimates that IGCC deployment would diminish a 500 MW plant's net capacity by 8% and increase the cost of electricity by 45%, while decreasing the related carbon emissions (kg CO2/kWh) by 89%.[37] The lack of commercial-scale CCS deployment limits the reliability of these data; however, current research makes it clear that CCS technology will not provide a simple or cheap solution to China's coal-carbon quandary.

3.3.4. Coal-Bed and Coal-Mine Methane

Effective capture of coal-bed and coal-mine methane presents an opportunity for increased energy productivity and safety in China's coal industry. However, methane capture has not been extensively developed in China due to lack of transparency regarding resource property rights and lack of available technology. These constraints are directly related to the consolidation of mine ownership and available capital in China.

On March 28, 2008, Shanxi Electric Power Corp. officially commenced operation of the world's largest CBM (coal-bed methane) power plant. In order to fire the plant's 120-MW capacity, the plant will consume 178.7 million cubic meters of methane and produce 840 GWh of electricity per year. In 2007, the Shanxi Jincheng Anthracite Coal Mining Group, the project's developer and one of the provinces largest CBM companies, produced 208.38 million cubic meters of CBM aboveground and 330.42 million cubic meters underground.[38]

CBM development was historically under the China United Coal-bed Methane Corporation (CUCBM) established in 1996. Because of the slow pace of development and consistent undershooting of production targets, CUCBM lost its monopoly rights in late 2007 as the government moved to open up the sector to further foreign investment. Indeed, CBM development is now the main focus of China's Clean Development Mechanism (CDM) projects, attracting 63% of the total UN-certified carbon credits in 2006.[39]

3.3.5. Mandatory Closure of Small-Scale Power Plants

In 2007, China shut down 553 small-scale generators with a total capacity of 14,380 MW—exceeding the central government's target by 43 percent. Companies that closed small generators in 2007 were promised government compensation equal to three years worth of revenue. Closures have also been facilitated by an expanded power generation quota trading program whereby small power plants are allowed to sell their production quotas to larger power plants and raise cash to manage shutdowns. Quota trading amounted to 53.6 billion kWh in 2007, which is estimated to have saved 6.16 mtce and 143,000 tonnes of sulfur dioxide emissions.[40]

3.3.6. Demand Scenarios

In order to quantify the potential savings of high-efficiency thermal power generation, this section examines coal requirements for the technology scenarios discussed in Section 3.3.2 above. At 2007 heat rates, China would require 2.8 billion tonnes of coal for electricity generation in 2025 (as illustrated by the dotted line in Figure 15). If China were able to improve its fleet efficiency from the baseline path to the hypothetical IGCC path, 14% less coal (349 million tonnes) would be required for electricity generation in 2025. Figure 15 shows that the baseline development path includes efficiency improvements that would require 269 million tonnes less coal in 2025 than generation requirements at the frozen 2007 average efficiency level.

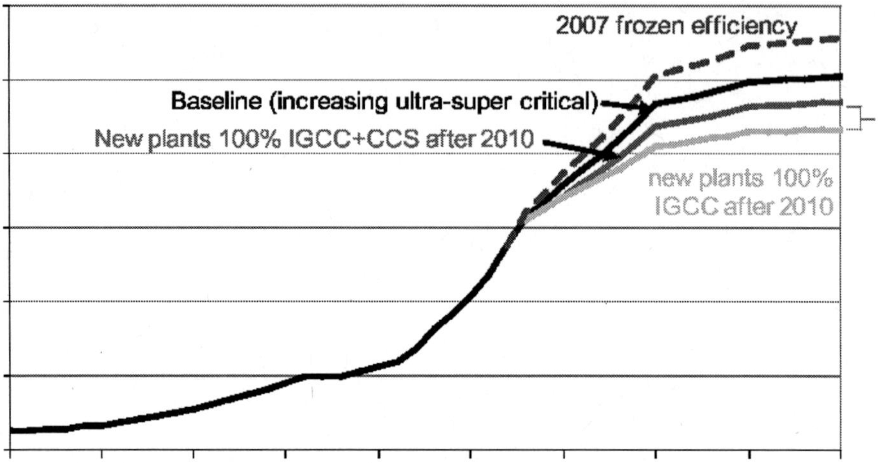

Figure 15. Coal Demand for Electricity Generation in Four Technology Scenarios, 1980-2025.

The improvement between the baseline and IGCC scenarios demonstrates that currently available thermal power generation technology does not offer a silver bullet for mitigating, much less reversing, rapidly growing thermal coal demand. The rapid build of power generation capacity between 2000 and 2007 will influence China's coal consumption path for many years.

3.4. Coking Coal for Iron & Steel Production

China has emerged as the largest producer and consumer of steel in the world. However, per-capita consumption remains below other industrialized countries. Table 2 shows the range of aggregate and per-capita steel consumption among a range of developing and developed countries. In 2008, China's per-capita steel consumption is expected to have reached 390 kg, thereby surpassing the US level in 2005.

Projections of future structural steel production are scaled off of building construction growth; historically, half of Chinese steel use was structural and the other half was finished products.[42] Product steel projections are based on the growth of other industrial sub-sectors. The coal requirement for steel production is based on assumptions regarding the share of primary (basic oxygen furnace) steel to secondary (electric arc furnace) steel and a constant share of steel exports. Based on urban population growth, building construction, and demand for finished products, Chinese crude steel production would grow from 489 million tonnes in 2007 to 588 million tonnes in 2025.[43] Increased recycling and more efficient primary production would reduce coking coal requirement from 416 million tonnes in 2007 to 291 million tonnes in 2025.

Table 2. Aggregate and Per-Capita Steel Consumption (2005)[41]

Country	Million tonnes	Kg per capita
China	350 mt	270 kg
United States	113 mt	382 kg
United Arab Emirates	-	1,314 kg
Taiwan	-	1,044 kg
Japan	83 mt	649 kg
India	41 mt	38 kg
World	1,013 mt	189 kg

3.5. Cement and Building Materials

In this analysis, future cement production levels are derived based on the amount of cement consumed to construct China's urban and rural buildings and infrastructure (roads, highways, bridges, airports, etc.). This methodology takes into account changes in the share of urban construction in total construction over time as well as changes in the average amount of cement used per square meter of construction. Cement production is driven by urbanization and the level of exports. The energy intensity of cement production varies by the assumed share of inefficient vertical shaft kilns as well as the timing of China reaching world best practice efficiency.

Annual urban construction and cement use data are combined with cement intensity to calculate total cement demand (total cement = (annual urban construction * cement intensity) / urban share of total cement). With reference to the growing average height of Chinese urban buildings, cement intensity of construction is assumed to grow from 220 kilograms of cement per square meter of new construction in 2007 to 260 kg/m^2 in 2025. Based on the assumptions and methodology described above, cement production is expected to decline from 1.36 billion tonnes in 2007 to 1.15 billion tonnes in 2015 and 1.04 billion tonnes in 2025; the coal requirement for non-metal mineral product manufacturing is also expected to drop from 170 million tonnes in 2007 to 120 million tonnes in 2025.

3.6. Coal-to-Liquids and Coal-to-Chemicals

Coal liquefaction and chemicals production present a potential fourth wedge of new coal demand between 2008 and 2025. To examine the potential impact of this burgeoning industry, this study calculates the coal feedstock requirements for 60 million tonnes of CTL capacity and 70 million tonnes per annum of coal-to-chemicals capacity in 2025.

In 2008, Shenhua coal company commissioned a 1 million-tonne direct coal liquefaction plant in Inner Mongolia.[44] China has reviewed plans for total coal-to-liquids capacity development of up to sixty million tonnes by 2020. At present, coal liquefaction requires 4.5 to 5 tonnes of coal and 10 tonnes of water per tonne of product.[45]

Regarding coal-to-chemicals, methanol production has displaced a growing share of ethanol production from grains and other feedstocks. Coal is increasingly viewed as a chemical feedstock. In 2007 China consumed 9 mt of methanol, of which 2.7 mt was blended with gasoline. Methanol is a feedstock for the production of dimethyl ether (DME) which is being piloted as a diesel fuel substitute in buses in Shanghai. By 2010 production capacity is expected to reach 39 mt of methanol. According to current technology, coal-to-chemicals production requires 1.5 to 1.7 tonnes of coal per tonne of methanol and 2 to 2.5 tonnes coal per tonne DME.

Based on current process efficiencies, growth to 60 million tonnes of CTL capacity and 70 million tonnes per annum of coal-to-chemicals capacity would require up to an additional 450 million tonnes of coal in 2025.

3.7. Demand Outlook Scenario

According to the development outlook of the basic drivers described above, Chinas total coal demand would grow to 4.26 billion tonnes in 2025 including coal-to-chemicals and coal liquefaction development, or 3.81 billion tonnes without CTL and CTC. In either case, electricity generation remains the largest driver of new coal demand. The growth of coal-tochemicals and coal liquefaction to 450 million tonnes of annual coal use in 2025 does not offset the growth of electricity generation from 50% of total use in 2006 to 59% in 2025. Figure 16 illustrates the dominant role of electricity generation (baseline scenario) in driving Chinese coal demand.

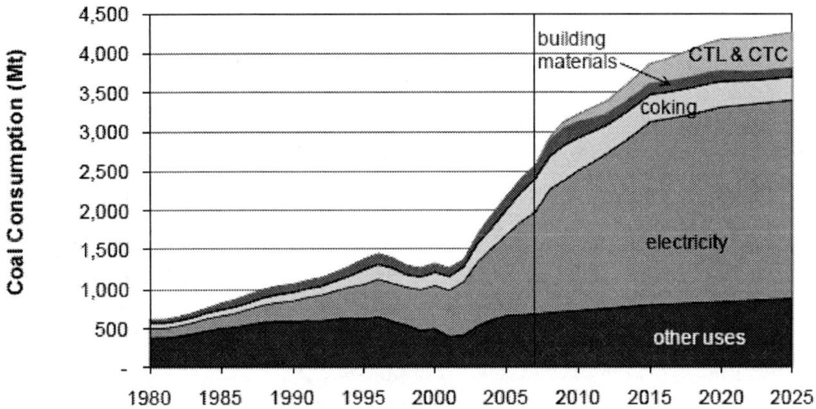

Figure 16. Historical and Forecast Coal Demand, 1980-2025.

Growth in the "other uses" category of Figure 16 is primarily driven by four energy trends in China. Residential final coal consumption declined from 1996 to 2006, but rebounded in 2006; unless residential natural gas prices rise dramatically, expanded residential gas use is expected to limit further coal growth. As district heating expands with urbanization in the northern heating zone, coal use for district heating will depend on the availability of natural gas as a preferential fuel. Fertilizer production will also drive coal demand growth as coal remains the primary Chinese feedstock and self-sufficiency in domestic food production remains a key state policy goal. Likewise other industry is expected to drive coal demand growth as few options exist for substitution; efficiency improvements are expected to moderate, but not reverse, demand growth.

IV. Coal Supply Constraints, Substitution Options, and Externalities

The number of published forecasts for Chinese coal production is limited. On the near-term end of the spectrum, coal production targets of the Tenth and

Eleventh Five-Year Plans were exceeded shortly after their publication. In the long term, the China Academy of Sciences has targeted zero growth of energy consumption after 2040.[46] Between China's official short and long-term projections, this study uses modified logistics curve analysis to quantify extraction scenarios in accord with published reserve data; these scenarios provide a foundation for analyzing coal gap potential between 2008 and 2025.

4.1. Constraints to Expanded Supply

In order to minimize the negative impacts of its potential coal gap, China is seeking to maximize domestic coal production. However, the rapid depletion of China's accessible, high-quality coal reserves is likely to hasten the arrival of peak production and accelerate post-peak decline. Aside from the larger resource depletion and strategy issues, China faces immediate practical constraints to further expanding its coal supply. Three key constraints to rapidly expanded Chinese coal production are freight transport bottlenecks, limits on electricity grid capacity to transmit coal by wire, and the diminishing quality of China's remaining coal reserves.

4.1.1. Transport Bottlenecks

As illustrated in Map 1, most of China's coal resources are located in the inland northern provinces of Shanxi, Shaanxi, and Inner Mongolia—away from coastal demand centers. Moving coal around the country utilizes a large and growing share of domestic transport capacity. In 2006 it is estimated that 80% of consumed coal was transported by rail, road, or water. Figure 17 shows that the estimated rail share of total transported coal surged to 75% in 2002 before receding to 60% in 2006.

In 2006, 1.1 billion tonnes of coal were transported by rail, an estimated 407 million tonnes by road, and 385 million tonnes by inland and coastal waterways. Although rail is the dominant mode of coal transport in China, it is more expensive and complicated than rail transport in other countries. National west-east rail freight costs for coal, for example, are 0.12 RMB/tonne-km ($0.017/tonne-km).[47] Furthermore, the scarcity of dedicated local rail lines means that coal must often be shipped on local rail links from mines to the national network at roughly double the national per tonne-kilometer rate. Because China lacks an equivalent to the interstate commerce clause of the U.S. constitution, trans-provincial shipments can be taxed several times before reaching their destination.

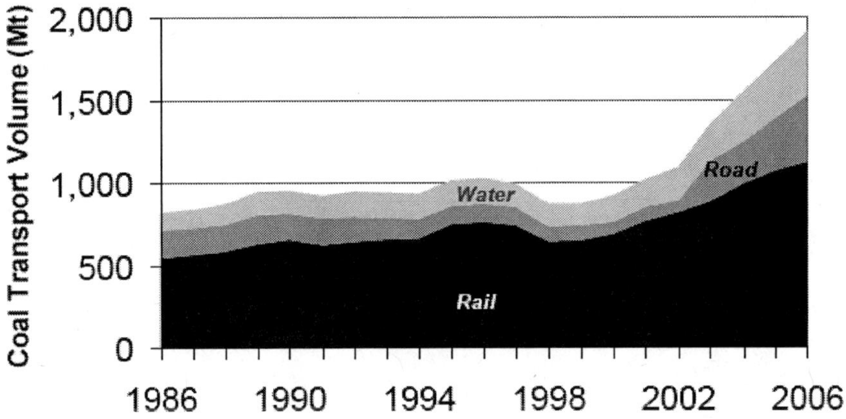

Figure 17. Coal Transport by Mode, 1986-2006.

Water provides an efficient mode of coal transport. However, rapid expansion of water transport capacity is limited by use of small and handysize vessels for domestic transport, lack of dedicated domestic coal port facilities, and—most of all—by lack of adequate rail capacity to bring coal to ports. When rail and water transportation is impossible, coal is transported by road. Road is the most expensive mode of coal transport in China. Fees range from 0.5 to 0.8 RMB per tonne-kilometer, but still do not curb demand for journeys of up to 300 km by truck.[48] To maximize transport volume and minimize costs, coal trucks—usually in the 20-tonne scale--are often overloaded, exceeding the design capacity of roads and highways and accelerating road deterioration. As with local rail freight, road transport fees are driven up by local and provincial taxes in China. Road, water, and particularly rail transport of coal already constrains current coal production and is likely to limit the rapid expansion of domestic supplies.

Coal transport comprised 46% of total rail freight in 2006. In anticipation of further rail freight transport growth, the government has announced a plan to expand the total commercial length of the national rail network from 77,000 kilometers in 2006 to 100,000 kilometers in 2020.[49] The national government's ambitious rail expansion targets appear to be insufficient to support the rapid growth of coal demand without further displacing other freight from the rail system. Figure 18 shows that growth of total rail freight and coal freight by rail has not kept up with coal consumption growth since 2004. Whereas coal use expanded 81% between 2000 and 2006, coal by rail and total rail freight only grew 63% and 47% respectively.

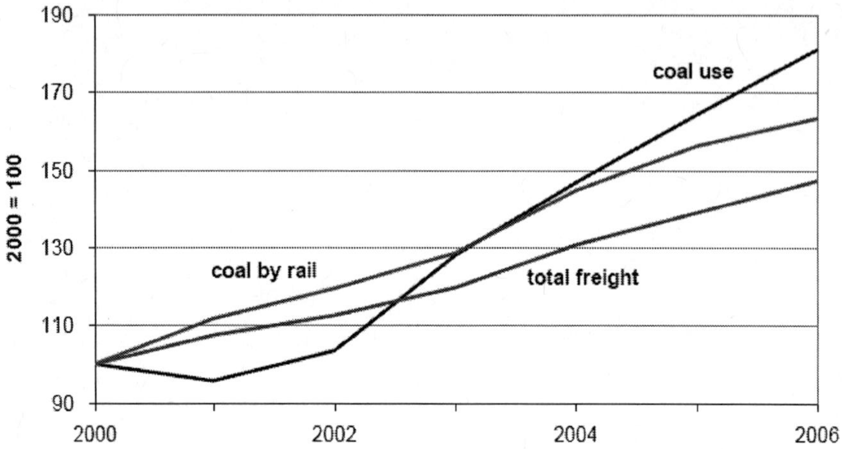

Figure 18. Index of China Coal Use, Coal Transport by Rail, and Total Rail Freight, 2000-2006.

Aside from bottlenecks within China's rail freight sector, coal transport is complicated by the geography and economic structure of the coal mining industry. While most of the key state-owned mines are served by rail lines, less than half the volume of 2006 coal was produced in rail-connected mines. Short-haul trucks are likely to have been used to move the 770 million tonnes of coal produced in near-to-rail mines in 2006. Road transport is likely to have been the only option for outside-of-the-system and difficult-transport mines, which comprised 17% of production in 2006. The prevalence of small coal mines distributed in mountainous terrain complicates efforts to efficiently transport more Chinese coal to markets.

Strong demand for coal ensures that mined reserves are moved to market, by whatever mode is available. In many cases in China, this means that insufficient or unavailable rail freight capacity is supplemented by overloaded trucks. Not only are road fees on average five times higher per tonne-kilometer than rail fees, but China's trucks are hugely energy inefficient in contrast to the rail system, which rivals Japan's in terms of energy efficiency. Figure 19 shows the estimated energy consumption (in thousand barrels diesel per day, hereafter abbreviated as kb/d) by mode from transporting coal in 2006 on the left and hypothetical consumption on the right, were all of the all coal transported by road actually transported by rail. Whereas 2006 actual coal transport consumed an estimated 116 kb/d (of which 67 kb/d was used for road transport), rail and water transport of the same amount of coal would have required just 54 kb/d. Indeed, the increasing growth of truck-transported, in addition to the shift to truck transport of other goods that have been displaced by the growing proportion of coal on the rail system, has been a major driver for the increase in transport fuels in recent years.

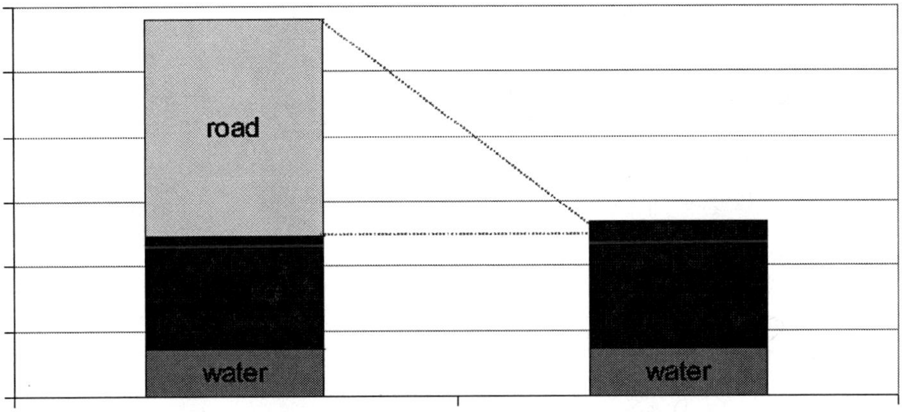

Figure 19. Actual and Hypothetical Bi-Modal China Coal Transport, 2006.

4.1.2. Grid Options

China's divergent distribution of coal resources presents attractive policy potential for increasing the use of coal by wire, whereby energy from coal is extracted by combustion at mine-mouth thermal power plants and then delivered long-distance to demand centers. Given China's current electricity grid and resource endowments, coal by wire presents significant short-term costs for a potential long-term solution.

The major benefit of "coal by wire" is the potential reduction of China's coal transport requirements. In addition to allowing other freight to be moved by rail, coal by wire could reduce oil demand by stemming coal road transport. However, coal by wire also has energy losses associated with transmission line losses in delivering the electricity. In China's case, these transmission and distribution losses were reported at 7.04% in 2006, compared to 5.85% for the US in 2006.[50] Another major challenge to coal by wire in China is the high water requirement of mine-mouth thermal power plants, which can reach 300 million gallons of water per day for a thermal power plant with 500 MW installed capacity.[51] This is a particularly significant disadvantage for China given the existing water shortage concerns in the Northern and Western regions.

4.1.3. Coal Quality: Lignite Substitution Scenario

Due to its low energy content, lignite is not as widely used as bituminous or anthracite coal. Although lignite composes 16% of China's coal reserves by mass, only 5% of the volume of China's 2005 coal production was lignite (see Figure 3). As more energy-rich and easily-accessible forms of coal are exhausted, lignite is likely to gain a growing share of national production. However, because the energy density of lignite is lower than that of bituminous or anthracite coal, a greater volume of lignite must be produced in order to deliver the same amount of energy contained in a lower volume of bituminous coal. Under conditions whereby the share of lignite production rises from its current 5% of total production to 12% of the total by 2025, an additional 220 million tonnes of additional coal would have to be mined to compensate for the lower energy content of the lignite (Figure 20). The energy content of U.S. coal production has also been declining: if 2006 coal had the same unit energy content as coal in 2000, the U.S. could have produced and transported 32 million tonnes less coal in 2006.[52]

The transition to lower energy density coal types will put further stress on rail transport and power generation systems. Coal energy density could be increased by 20%--thus reducing the transport pressures—with more extensive coal washing. However, the high water requirement for coal washing, at 4 tonnes per tonne of raw coal, has constrained expansion of washing capacity in the water-deficit regions of North China.

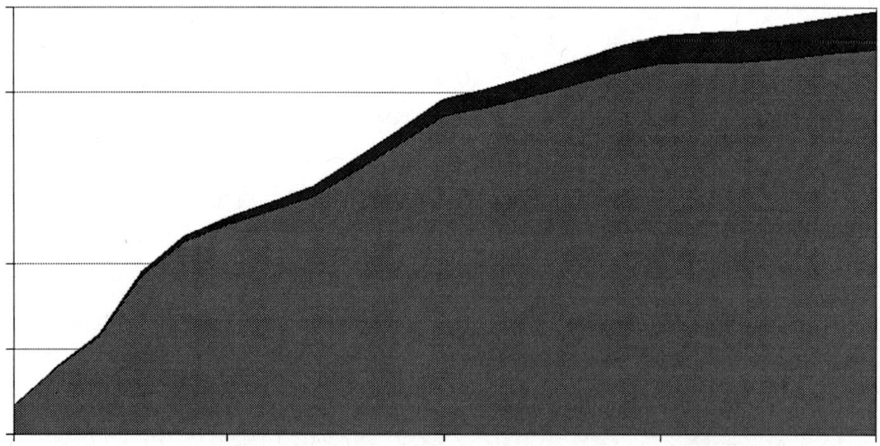

Figure 20. Additional Production Requirements with 12% Lignite, 2005-2025.

4.2. Demand-Supply Gap under Three Scenarios

The mid-term future of Chinese coal production serves as the physical foundation for economic growth, political stability, and ballooning carbon emissions. Given the Ministry of Land and Resources estimate of 189 billion tonnes remaining domestic coal reserves and the historical production trajectory of other coal-intensive countries such as the United Kingdom, Japan, and Germany, Chinese production is likely to peak in the foreseeable future. The shape and timing of China's coal production peak will depend on technology, investment, and the water availability and geology of new discoveries. This analysis uses modified logistics curve analysis to indicate future production levels according to the Ministry of Land and Resources indicated economic reserve estimate of 189 billion tonnes coal. Three production scenarios are developed: a sharp peak whereby growth and decline are accelerated by aggressive investment, effective technology, and available reserves; a broad peak with more gradual extraction; and a later peak as an additional 30% of the resource base is converted into economically exploitable reserves. Figure 21 juxtaposes these three coal production scenarios with the total demand data illustrated in Figure 16.

Whereas logistic analysis of reserve data generates 2025 domestic production between 3.25 billion tonnes for a broad peak scenario and 4.16 billion tonnes for a sharp peak scenario, the IEA has forecast 2025 Chinese coal production of approximately 4.33 billion tonnes and Chinese academic analysis posits production of 3.72 billion tonnes in 2025.[53] On the basis of published reserve data, none of the LBNL domestic coal scenarios include a production peak before 2025. If coal demand grows as described in Figure 16 (including coal-to-chemicals and coal liquefaction programs, as illustrated in the dotted line), these production scenarios would lead to a 2025 domestic coal gap between 32 million tonnes and 990 million tonnes. The scale of China's potential 2025 coal gap overshadows international coal imports, which reached an historic high of 51 million tonnes in 2007. A sharp peak production scenario would bring coal supply closer to demand; however, it would also require a more rapid draw down of existing reserves and a potentially more rapid post-peak decline.

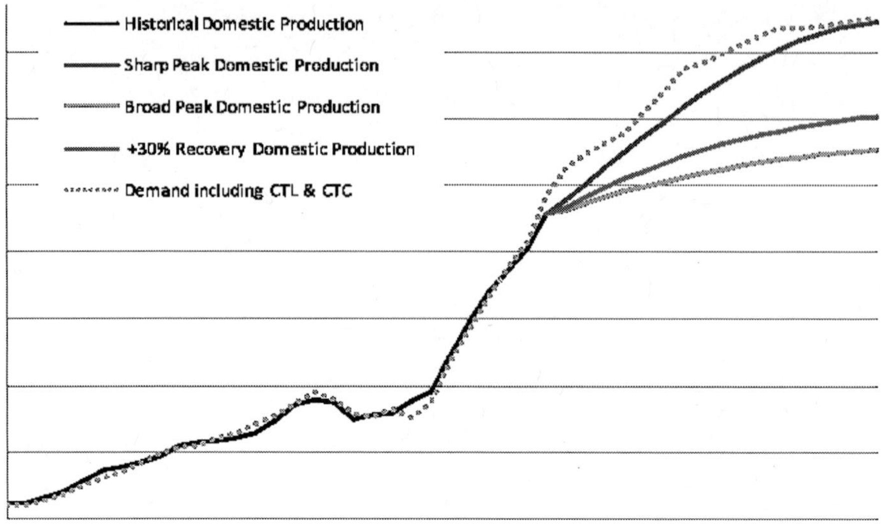

Figure 21. China Coal Production Scenarios, 1980-2025.

4.3. Fuel Switching

One Chinese response to the specter of coal deficits is to maximize the growth of alternative and especially renewable energy generation. In 2006, China implemented a Renewable Energy Law that aims to boost China's renewable energy capacity to 15 percent of total by the year 2020 and outlines a commitment to invest $180 billion in renewable energy over that period. However, the scale of coal demand growth in China—between 2003 and 2007 coal consumption growth ranged between 190 and 330 million tonnes per year—presents a challenge for fuel substitution.

Natural Gas

Idle natural gas-fired electricity capacity estimated at 10 GW out of 15 GW total indicates that China is already suffering shortages of supply despite a rapid rise in production over the past 5 years. If natural gas were used to offset one year of coal annual demand growth (approximately 200 million tonnes), 107 billion cubic meters of additional natural gas would be required on a thermal equivalent basis (for end use) or 83 billion cubic meters would be needed on a generation equivalent basis, assuming 63 GW of new capacity and a 70% load factor.

Table 3. Energy Requirements for offsetting 200 Million Tonnes of Coal per year (~ one year of growth)

Thermal Equivalent	Generation Equivalent
107 billion cubic meters of natural gas	83 billion cubic meters of natural gas and 63 GW of new capacity @ 70% load factor
	48 GW of nuclear power capacity @ 90% load factor
	86 GW of hydropower capacity @ 50% load factor
	143 GW of wind power capacity @ 30% load factor

This additional volume is beyond China's capacity to supply within the next 10 years, and would require the construction of at least two world-scale pipelines of 30 billion cubic meter capacity each as well as a faster ramp-up of LNG terminal construction and the realization of additional supply contracts. In additional, as long as electricity prices remain controlled at levels commensurate with overall coal pricing, natural gas is unlikely to be competitive against coal except in specific circumstances and locations, such as in southern coastal China or near LNG import terminals.

Nuclear Power

China seeks to rapidly develop its nuclear power industry primarily to increase energy supply. The complexity, social risk, and high capital requirements are further justified by the nuclear power industry's potential to help mitigate problems from uneven resource distribution, coal transport bottlenecks, and environmental pollution. With coal and liquid transport fuel deficits in mind, Chinese researchers have suggested that nuclear power could be used for electricity generation, thereby freeing up coal for liquefaction. A more ambitious goal is to develop nuclear-coal conversion technology to produce liquid transport fuels based on high temperature nuclear process heat.[54]

In order to offset one year of coal demand growth of approximately 200 million tonnes coal, 48 GW of new nuclear power capacity would be required, assuming a 90% load factor for all new plants. At a capacity of 1 GW and a conservative cost of $1.5 billion per reactor, nuclear substitution of coal would cost $72 billion for the core hardware. Given China's announced 2020 target total nuclear electricity generation capacity target of 40 GW, it would take more than a decade to offset one year of coal demand growth.

China's nuclear power capacity reached 9.07 GW in 2007. Although the government has released aggressive plans for nuclear power development, it has not explained how it will ensure adequate domestic uranium supplies. An installed nuclear power capacity of 40 GW would require approximately 7,000 tonnes of uranium on an annual basis.[55] At a price of at least $130 per kilogram of uranium, China's identified resources are estimated at 60,000 tons of uranium.[56]

However, China has increasingly close trading relationships with Australia and Kazakhstan, which is estimated to have the world's second largest uranium resources at 816,000 tons. In 2006 China signed an agreement to purchase 20,000 tonnes of Australian uranium per year beginning in 2010.[57] Nevertheless, the growth of nuclear power throughout Asia suggests that China's reliance on imported uranium may constrain rapid large-scale development.

Hydropower and Renewables

Without effective battery or storage technology, hydropower and renewables are imperfect substitutes for coal due to their intermittency. China is seeking to double its hydropower capacity from 145 GW in 2007 to 300 GW. In order to substitute for one year of new coal demand growth, 86 GW of new hydropower capacity would be required, assuming a 50% load factor. This is equivalent to building almost four Three Gorges Dams to offset one year of coal demand growth.

In 2007, wind power capacity jumped to 6 GW; however, wind still comprised less than 1% of China's total electricity generation capacity. Assuming a 30% load factor, 143 GW of new wind power capacity would be required to offset one year of coal demand growth. In 2008 China announced a new 100 GW target for wind electricity generation capacity in 2020—from its 2007 base, this would require an average annual growth rate of 24%.

4.4. Environmental Externalities

Coal in China can be characterized as a cheap and abundant source of energy with expensive externalities. Coal mining and combustion are associated with a range of environmental costs including land subsidence, degeneration of water quality, air pollutant emissions, and acid rain, not to mention the horrific human costs related to more than 5,000 miner deaths per year. At the same time, the coal industry serves as a major foundation of rural employment and urban growth though electricity production. Beyond the immediate human and environmental costs, greenhouse gas emission mitigation provides an international impetus for China to tackle its coal dependence.

4.4.1. Carbon Dioxide Emissions

Coal is China's most carbon-intensive primary energy source. According to the IPCC, coal generates an average 95 tonnes of CO2 emissions per terajoule (TJ) of energy, compared to 73 t CO2/TJ for oil and 56 t CO2/TJ of natural gas.[58] The coal share of China's energy-related carbon emissions increased from 78% to 80% between 2000 and 2006. China's increasing reliance on coal is reflected in the rise of carbon intensity of commercial energy production from 2.26 kg carbon dioxide emitted per kg of coal equivalent energy produced in 1985 to 2.7 kg in 2006. The root of China's rising carbon emissions is clear: in 2006, coal combustion accounted for 80% of energy-related Chinese carbon dioxide emissions.

China's current growth trajectory indicates that carbon emissions from coal combustion will surpass total US energy-related carbon emissions by 2010. Figure 22 shows that China's coal-related carbon emissions would surge to 8.6 billion tonnes of CO2 in 2025 under the demand trajectory developed in this report (including CTL and CTC; 7.7 billion tonnes without the fourth wedge).[59] In contrast, the WEO 2007 projects total US energy-related carbon dioxide emissions to reach 6.7 billion tonnes in 2025.

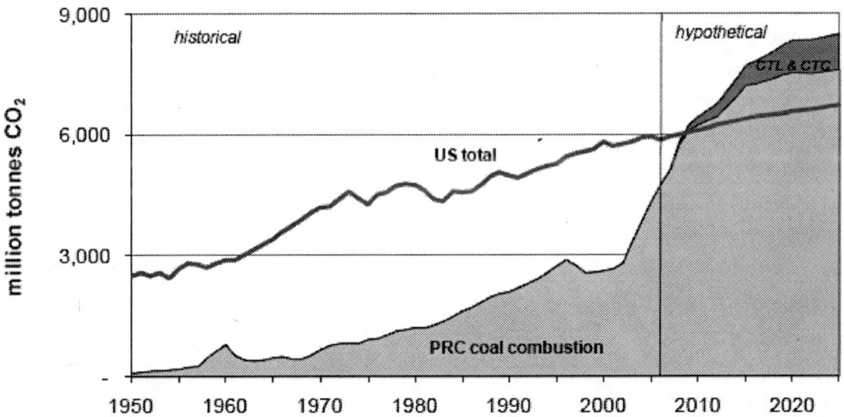

Note: historical and forecasted US emission data are from the IEA World Energy Outlook, 2007.

Figure 22. China Coal-related Carbon Dioxide Emissions, 1950-2030.

Limited alternatives to coal impede China's flexibility to achieve carbon dioxide emission growth constraints, much less absolute reductions.

4.4.2. Environmental Quality

Coal mining and combustion are associated with a range of environmental costs including land subsidence, degeneration of water quality, air pollutant emissions, and acid rain. Sulfur dioxide and particulates are considered by many environmental experts in China to be the air pollutants 0 of gravest concern, and efforts at controlling air pollution have focused on them (Figure 23). Ash and sulfur levels vary among Chinese coal types: raw ash fluctuates between 20% and 40% and sulfur between 1% and more than 5%. Since the 1980s, the fraction of China's coal that has been washed has been stable, and flue gas desulfurization is only now becoming widespread.

As more becomes known about the health impacts of airborne particulates, especially very small particulates that are drawn deep into the lungs, they are considered responsible for a greater share of the world's ill health than previously thought (WHO, 2004). Particulate emissions from combustion and physical processes (like industrial grinding) fell rapidly in China in the 1990s as relatively inexpensive particulate controls were installed on a larger share of industrial facilities. In recent years, however, rising coal use and industrial activity have led to an upturn in particulate emissions.

The impacts of a coal-dominated energy economy are felt all along the fuel chain. About 4,000 miners die each year in accidents in China's coal mines, according to official statistics, though the actual number may be substantially higher. An unknown number also die from occupational diseases. Coal gas and coking plants are also associated with high incidences of cancer. While new combustion technologies and improved regulation will help to abate some of the most egregious health impacts, the ongoing dominance of coal portends continued environmental costs associated with expanded energy usage.

Nationwide damages from air and water pollution have been estimated to have a value between 3% and 7.7% of GDP. The destructive impact of air pollution on human health and physical infrastructure is likely to increase as rapid urbanization and expanding cities locate a growing proportion of the population nearer to pollution sources.[60]

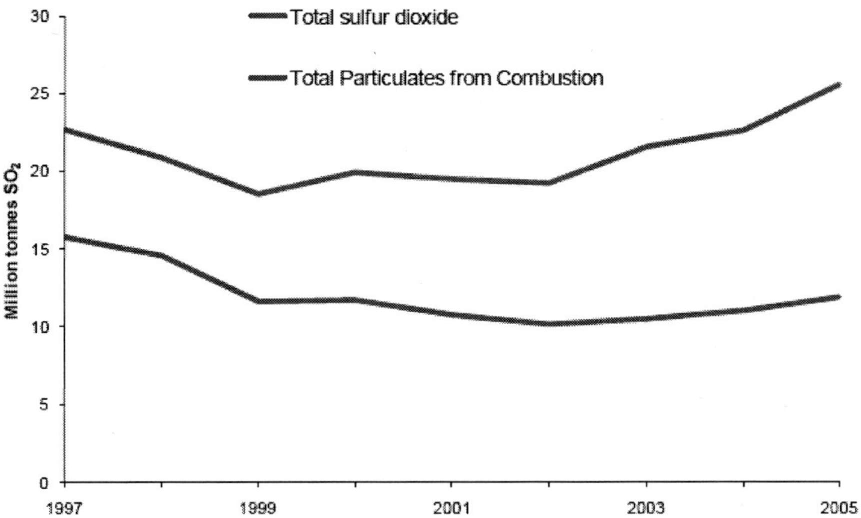

Source: China Environment Yearbook, various years.

Figure 23. Sulfur dioxide and particulate emissions, 1997-2005.

CONCLUSIONS

China has been, is, and will continue to be a coal-powered economy. The rapid acceleration of coal demand growth after 2001 has raised questions about China's mid-term supply security. If demand continues to grow rapidly and supply develops according to published domestic reserve estimates a coal deficit is likely to emerge.

Urbanization, heavy industry growth, and increasing per-capita consumption are the primary drivers of rising coal usage. In 2006, the power sector, iron and steel, and cement accounted for 71% of coal consumption. Power generation is becoming more efficient, but even extensive rollout of the highest efficiency units could save only 14% of projected 2025 coal demand. A new wedge of future coal consumption is likely to come from the burgeoning coal-liquefaction and chemicals industries. New demand from coal-to-liquids and coal-to-chemicals may add up to 450 million tonnes of coal demand by 2025. Growth with efficiency improvements among these drivers indicates that China's annual coal demand will reach 3.9 to 4.3 billion tonnes by 2025.

Central government support for nuclear and renewable energy has not been able to reduce China's growing dependence on coal for primary energy. Few substitution options exist: offsetting one year of recent coal demand growth would require over 107 billion cubic meters of natural gas, 48 GW of nuclear, or 86 GW of hydropower capacity. While these alternatives will continue to grow, the scale of development using existing technologies will be insufficient to substitute significant coal demand before 2025. The central role of heavy industry in GDP growth and the difficulty of substituting other fuels suggest that coal consumption is inextricably entwined with China's economy in its current mode of growth.

Ongoing dependence on coal reduces China's ability to mitigate carbon dioxide emissions growth. If coal demand remains on its current growth path, carbon dioxide emissions from coal combustion alone would exceed total US energy-related carbon emissions by 2010. Broadening awareness of the environmental costs of coal mining, transport, and combustion is raising the pressure on Chinese policy makers to find alternative energy sources.

The looming coal gap threatens to derail China's growth path, possibly undermining political, economic, and social stability. High coal prices and domestic shortages will have regional and global effects. Regarding China's role as a global manufacturing center, a domestic coal gap will increase prices and constrain growth. Within the Asia-Pacific region, China's coal gap may bring about increased competition with other coal-importing countries including Japan, South Korea, Taiwan, and India. As with petroleum, China could respond with a government-supported "going-out" strategy of resource acquisition and vertical integration. Given its population and growing resource constraints, it may be difficult for China to balance energy security, competitiveness, and local environmental protection with costly efforts at global climate change mitigation. The possibility of a large coal gap suggests that Chinese and international policy makers should maximize institutional and financial support to moderate demand and improve energy efficiency.

REFERENCES

Aden NT, Sinton JE. 2006. "Environmental implications of energy policy in China," *Environmental Politics.* 15(2):248-70.

Andrews-Speed, P. (2004) *Energy Policy and Regulation in the People's Republic of China* (The Hague: Kluwer Law International).

Chen C; Rubin ES. 2009. "CO2 control technology effects on IGCC plant performance and cost," *Energy Policy* 37: 915-924.

China State Electricity Regulatory Commission (SERC). 2009. http://www.serc.gov.cn/jgyj/ztbg/200903/W020090324593421835268.pdf (May 2009).

Development Research Center (2004) National Energy Strategies and Policies Report (Beijing: Development Research Center). http://www.efchina.org/documents/Draft Natl E Plan0311.pdf (March 2005).

Energy Watch Group (EWG). 2007. *Coal: Resources and Future Production.* Berlin: EWG-Paper No.1/07. http://www.energywatchgroup.org/fileadmin/global/pdf/EWG Report Coal 10-07-2007ms.pdf

Finkelman, R.; Beklin, H. and Zheng, B. (1999) 'Health impacts of domestic coal use in China,' *Proceedings of the National Academy of Science* 96: 3427-3431.

Gerbens-Leenes, Winnie and Hoekstra, Arjen and Meer van der, Theo (2007) "The Water Footprint of Energy Consumption: an Assessment of Water Requirements of Primary Energy Carriers." *ISESCO Science and Technology Vision,* 4 (5). pp. 38-42.

Gibbins, Jon; Chalmers, Hannah. 2008. "Carbon Capture and Storage," *Energy Policy* 36: 4317-4322.

Glomsrod, S. and Wei T. (2005) 'Coal Cleaning: a viable strategy for reduced carbon emissions and improved environment in China?' *Energy Policy* 33, pp. 525-542.

Guan Dabo, Klaus Hubacek. (2007) "Assessment of regional trade and virtual water flows in China," *Ecological Economics* 61:159-170.

IEA 2008, various years. *World Energy Outlook.* Paris: OECD/International Energy Agency. IEA. 2009. *Cleaner Coal in China.* Paris: OECD/International Energy Agency.

Levine Mark, Aden Nathaniel. 2008. "Global Carbon Emissions in the Coming Decades: The Case of China," *Annual Review of Environment and Resources*, Vol. 33: 19-38.

Li Zhenzhong, et al. 2004. "Ultra-Supercritical Coal Combustion for Electricity Generation: Technology Choice and Industrial Development," China Association of Scientists, 2004. (*in Chinese*: "超超临界燃煤发电机组的技术选择与产业化发展")

Liu Hengwei; Ni Weidou; Li Zheng; Ma Linwei. (2008) "Strategic thinking on IGCC development in China," *Energy Policy* 36 (2008): 1-11.

Mi Jianhua. 2006. "Analysis of Energy Efficiency Status of Power Generation Industry in China," *Electrical Equipment*, 7(5): 9-12. (*in Chinese*)

National Comprehensive Energy Strategy and Policy (NESP). 2005. The Development Research Center of the State Council of China. May. Available at http://www.efchina.org/documents/Draft Natl E Plan0311.pdf (condensed version).

National Development and Planning Commission Department of Investment, National National Development Reform Commission (NDRC). 2007. *11th Five Year Plan on Energy Development*, NDRC, Beijing.

National Development Reform Commission (NDRC). 2007a. *11th Five Year Plan on Coal Industry Development*, NDRC, Beijing.

Nolan, P., Shipman, A., Rui, H. (2004) 'Coal Liquefaction, Shenhua Group, and China's Energy Security,' *European Management Journal* 22(2): 150-164.

Stern, Jonathan, Ed. (2008). *Natural Gas in Asia: The Challenges of Growth in China, India, Japan, and Korea, Second Edition*. Oxford: Oxford University Press.

Tao Zaipu; Li Mingyu. 2007. "What is the limit of Chinese coal supplies—A Stella model of Hubbert Peak," *Energy Policy* 35(2007): 3145-3154.

Thomson, E. (1996) 'Reforming China's Coal Industry,' *The China Quarterly* 147: 726-750.

United States Energy Information Administration (2006). *International Energy Outlook 2006*. Washington DC: EIA.

WEC 2007. World Energy Council, *Survey of Energy Resources,* 2007. Online at http://www.worldenergy.org/documents/ser2007 final online version 1.pdf; accessed May 2009.

Zhao LF, Xiao YH, Gallagher KS, Wang B, and Xu X. 2008. "Technical, environmental, and economic assessment of deploying advanced coal power in the Chinese context," *Energy Policy*, 36(2008): 2709- 2718.

End Notes

[1] World Energy Council. 2007 Survey of Energy Resources. London: WEC Press.

[2] MLR. 2003. *National Coal Resources and Reserves Circular*. Beijing: MLR, cited in IEA. 2009. *Cleaner Coal in China*. Paris: IEA.

[3] Ghee Peh, Wei Ouyang. (2007) *UBS Investment Research: China Coal Sector*. (17 January 2007; www.ubs.com/investmentresearch) ; NBS. 2007. *China Statistical Yearbook 2007*. Beijing: NBS Press.

[4] Data are sourced from the Ministry of Land and Natural Resources and NBS. Display sourced from http://fti.neep.wisc.edu/neep533/SPRING2004/lecture2.pdf.

[5] Ghee Peh, Wei Ouyang. (2007) *UBS Investment Research: China Coal Sector*. (17 January 2007; www.ubs.com/investmentresearch)

[6] World Energy Council. 2007 Survey of Energy Resources. London: WEC Press.

[7] IEA. (2007) *World Energy Outlook 2007*.

[8] Pan Kexi. (2005) "The Depth Distribution of Chinese Coal Resource," Fudan University, School of Social Development and Public Policy. August 22, 2005.

[9] IEA. (2007) *World Energy Outlook 2007*.

[10] World Energy Council. 2007 Survey of Energy Resources. London: WEC Press.

[11] World Energy Council. 2007 Survey of Energy Resources. London: WEC Press.

[12] International Energy Agency (IEA). 2007. *World Energy Outlook 2007*. Paris: IEA.

[13] National Development Reform Commission (NDRC). 2007. *11th Five Year Plan on Energy Development*, NDRC, Beijing. Separate media reports indicate that the actual recovery rate may be lower. One report, for example, says that China's average coal recovery rate is 30% nationally and 40% in Shanxi province (http://www.cnmn.com.cn/Show_26346.aspx).

[14] National Development Reform Commission (NDRC). 2007. *11th Five Year Plan on Energy Development*, NDRC, Beijing.

[15] See for example http://www.guardian

[16] Guan Dabo, Klaus Hubacek. (2007) "Assessment of regional trade and virtual water flows in China," *Ecological Economics* 61:159-170.

[17] Elspeth Thomson. (2003). *The Chinese Coal Industry*. New York: RoutledgeCurzon, 192.

[18] Gerbens-Leenes, Winnie and Hoekstra, Arjen and Meer van der, Theo (2007) *The Water Footprint of Energy Consumption : an Assessment of Water Requirements of Primary Energy Carriers.* ISESCO Science and Technology Vision, 4 (5). pp. 38-42.

[19] See for example http://news2.eastmoney.com/080527,1025,848560.html.

[20] Interfax. 2008. *China Energy Report Weekly*. Vol. 8 (18):23.

[21] Interfax. 2009. *China Energy Report Weekly*. Vol. 8 (6):18.

[22] World Energy Council. 2007 Survey of Energy Resources. London: WEC Press.

[23] United Nations. 2007. World Urbanization Prospects: The 2007 Revision Population Database (http://esa.un.org/unup/index.asp?panel=3; accessed May 5, 2008) McKinsey Global Institute. 2008. "Preparing for China's Urban Billion: Summary of Findings," Shanghai: McKinsey Global Institute, March 2008.

[24] *Interfax China Energy Report Weekly*, April 3, 2008, Vol. 7:14, p. 5.

[25] Lam, P., 2004, "Pricing of electricity in China," *Energy* 29 (2): 287-300.

[26] Wang, B., 2007, "An imbalanced development of coal and electricity industries in China," *Energy Policy* 35 (10): 4959-4968.

[27] Andrews-Speed, et. al, 1999, "Do the Power Sector Reforms in China Reflect the Interests of Consumers?" *The China Quarterly* 158: 430 – 446.

[28] Mi Jianhua. 2006. "Analysis of Energy Efficiency Status of Power Generation Industry in China," *Electrical Equipment*, 7(5): 9-12. (*in Chinese*)

[29] Li Zhenzhong, et al. 2004. "Ultra-Supercritical Coal Combustion for Electricity Generation: Technology Choice and Industrial Development," China Association for Science and Technology, 2004. (*in Chinese*: "超超临界燃煤发电机组的技术选择与产业化发展"); IEA, WEO 2007; IEA Greenhouse Gas Program. Lower efficiency levels have also been estimated for some of these technologies, depending on the specific circumstances of deployment.

[30] China Daily (July 2, 2007) "Being Supercritical,"

(http://www.zoomchina.com.cn/new/content/view/26601/266, accessed May 2008).

[31] Sun Guodong. (2008) *Coal in China: Resources, Use, and Advanced-Coal Technologies*. Pew Center on Global Climate Change. Mi (2006).

[32] China Daily (July 2, 2007) "Being Supercritical," (http://www.zoomchina.com.cn/new/content/view/26601/266, accessed May 2008).

[33] Average efficiency data for 2007 quoted in 全国电力工业统计快报 (2007) CEC.

[34] Liu Hengwei; Ni Weidou; Li Zheng; Ma Linwei. (2008) "Strategic thinking on IGCC development in China," *Energy Policy* 36 (2008)1-11.

[35] Sun Guodong. (2008) *Coal in China: Resources, Use, and Advanced-Coal Technologies*. Pew Center on Global Climate Change. Mi (2006).

[36] Gibbins, Jon; Chalmers, Hannah. 2008. "Carbon Capture and Storage," *Energy Policy* 36: 4317-4322. Chen, Chao; Rubin, Edward S. 2009. "CO2 control technology effects on IGCC plant performance and cost," *Energy Policy* 37: 915-924.

[37] *Ibid.*

[38] *Interfax China Energy Report Weekly*, March 27, 2008, Vol. 7:13, p. 13.

[39] Fridley, David, "Natural Gas in China", in Jonathan Stern, ed, *Natural Gas in Asia*, Oxford: Oxford University Press, 2008

[40] *Interfax China Energy Report Weekly*, March 27, 2008, Vol. 7:13, p. 8.

[41] Source: World Coal Institute. Coal & Steel Facts 2007; International Iron & Steel Institute.

[42] Urandaline Investments. 2009. "China's Infrastructure Boost and Its Impact on 2009 Metal Demand" Beijing: Access China Conference.

[43] NBS 2008 reported 2007 crude steel production of 489 million tonnes; personal communication with Chinese steel specialists suggests that China's 2007 crude steel production was actually 449 million tonnes.

[44] Media reports conflict as to whether the plant is located in Shanxi or Inner Mongolia. See http://www.platts.com/weblog joins south africa as a.html and http://www.bloomberg.com/apps/news?pid=20601072&sid=a3SIXbVZ8JiE&refer=energy

[45] Nolan, P., Shipman, A., Rui, H. (2004) 'Coal Liquefaction, Shenhua Group, and China's Energy Security,' *European Management Journal* 22(2): 150-164.

[46] People's Daily Online. 2007. "What will China be like in 2050?" (http://english.people.com.cn/200702/13/eng20070213 349784.html, accessed May 2008)

[47] IEA. 2007. WEO 2007. Paris: IEA.

[48] IEA. 2007. WEO 2007. Paris: IEA.

[49] Target announced in the 'Mid and Long-Term Railway Development Plan,' quoted in WB China Quarterly Update 2.2008.

[50] China State Electricity Regulatory Commission (SERC). 2009. http://www.serc.gov.cn/jgyj/ztbg/200903/W020090324593421835268.pdf.

[51] Feeley, T. et. al. 2005. "Department of Energy/Office of Fossil Energy's Power Plant Water Management R&D Program." www.netl.doe.gov/technologies/coalpower/ewr/pubs/IEP Power Plant Water R&D Final 1.pdf.

[52] U.S. EIA. 2008. 2007 Annual Energy Review. http://www.eia.doe.gov/aer/coal .

[53] IEA, World Energy Outlook 2007. Paris: IEA. Tao Zaipu; Li Mingyu. 2007. "What is the limit of Chinese coal supplies—A Stella model of Hubbert Peak," *Energy Policy* 35(2007): 3145-3154.

[54] Wang DZ, Ly YY. 2002. "Roles and prospect of nuclear power in China's energy supply strategy," *Nuclear Engineering and Design* 218(2002): 3-12.

[55] *Interfax China Energy Report Weekly*, March 27, 2008, Vol. 7:13, p. 4.

[56] World Energy Council. 2007 Survey of Energy Resources. London: WEC Press.

[57] See, for example, http://news.bbc.co.uk/2/hi/asia-pacific/4871000.stm.

[58] IPCC. 1996. Revised 1996 IPCC Guidelines for National Greenhouse Gas Inventories: Reference Manual, 1.13. (http://www.ipcc-nggip.iges.or.jp/public/gl/guidelin/ch1ref1.pdf)

[59] This emissions scenario does not include potential effects from carbon capture and storage programs.

[60] Ho, Mun S. & Chris P. Nielsen. (2007) *Clearing the Air: The Health and Economic Damages of Air Pollution in China.* Cambridge, MA: MIT Press.

INDEX

D

E

F

G

H

I

S

T